Decommissioning th
Spar

Decommissioning the Brent Spar

Tony Rice and Paula Owen

London and New York

E & FN Spon
An imprint of Routledge
London and New York

First published 1999
by E & FN Spon, an imprint of Routledge
11 New Fetter Lane, London EC4P 4EE

Simultaneously published in the USA and Canada
by Routledge
29 West 35th Street, New York, NY 10001

Typeset in Times New Roman by Routledge
Printed and bound in Great Britain by
TJ International Ltd, Padstow, Cornwall

British Library Cataloging in Publication Data
A catalogue record for this book is available from the British
Library

Library of Congress Cataloging in Publication Data
Decommissioning the Brent Spar/Tony Rice and Paula Owen.
Includes bibliographical references.
1. Oil storage tanks–Decommissioning–North Sea Region.
2. Oil storage tanks–Environmental aspects–North Sea Region.
3. Environmental policy–Great Britain.
4. Brent Spar (oil storage tank) I. Owen, Paula. II. Title.
TN871.4.R53   1999
363.738'2'0941–dc21   98–46095
                                        CIP

ISBN 0–419–24080–2 (hbk)
ISBN 0–419–24090–X (pbk)

Contents

Illustrations

Acknowledgements

This book started life as two separate works, one from the slightly detached point of view of an environmental information specialist, the other from that of a deep sea biologist who became embroiled in the public arguments surrounding the Brent Spar during the summer of 1995. Combining the two resulted in the loss of much of the detail in the originals. Both authors are grateful to a number of individuals and organisations for help in producing both what survived and what was cut out.

First, we must thank our respective partners for allowing us to devote so much quality time to 14,500 tonnes of inanimate metal. Second, we thank the two principal protagonists for their help, particularly Eric Faulds, Fran Morrison and their respective teams within Shell, and numerous Greenpeace personnel. They freely provided information as did many friends and colleagues in other organisations, including particularly the Scottish Association for Marine Sciences and the SOAEFD Marine Laboratory in Aberdeen, even though they knew that our approach might be somewhat critical. Our colleagues at SOC and the British Library, particularly librarians, also gave invaluable assistance and often moral support. Finally, we are especially grateful to Rod Smith, ex oil man, net surfer and provider of vast amounts of information that we would otherwise never have found. To all we extend our thanks, apologise for any failure to reflect adequately their contributions, and accept total responsibility for the interpretations and opinions expressed.

Introduction

Why write a book about the Brent Spar?

In the early summer of 1995, one of the largest oil companies in the world tried to get rid of the redundant 14,500-tonne Brent Spar by sinking it in the deep sea. A potent mixture of Greenpeace, a number of European ministers and an outraged public stopped them. During the six weeks between Greenpeace's initial occupation of the Spar and Shell's capitulation the story dominated the news in the UK, in Europe and around the world. It was to take Shell almost three years to find an acceptable alternative method of disposing of the Spar – its re-use in a quay extension in Norway. This was a glorious victory for the environment, common sense and the little man, and one in the eye for bureaucracy and the corporate power of big business. Or was it? Whatever else it was, it revealed serious deficiencies in the way these decisions are arrived at, at least in the UK. And it was a classic example of a monumental error of judgement on the part of Shell and the British Government, exploited highly successfully by Greenpeace. The whole episode has been surrounded by blunders, confusion and misinformation. This book attempts to explain what happened and why, and discusses the wider, far-reaching implications of the unique story that was 'Brent Spar'.

When the Spar burst onto the European news media on Sunday 30 April 1995, it took almost everyone, including Shell, completely by surprise. Hardly anyone outside the offshore oil industry – and Scotland – had even heard of the Spar, let alone knew what it was or what all the fuss was about. Those in the know in the industry were well aware of Shell's plans to dispose of it, because its fate could be very important for many other companies. And Scots, particularly in Aberdeen where most of the North Sea oil operations are based, had been fed regular items about the Spar by their local media for some months previously. Even so, the events of that late spring weekend pushed the Spar into a quite different ballgame. Suddenly, the headline slots on radio and television news and the front pages of national newspapers were dominated by dramatic pictures and descriptions of this curious, rusting, red-and-yellow object poking out of the waves somewhere in the wastes of the northern North Sea – and of the Greenpeace activists who had decided to hijack it. For seven weeks, until Mr Major's resignation and the ensuing Redwood–Major battle for the leadership of the governing Conservative party drove everything else from the headlines, the Spar became one of the main talking points both in the UK and continental Europe. It even sent shock waves around the world, to Japan, to Australasia, to South Africa and to North America.

There were, of course, 'dead' periods when nothing much seemed to be happening. But these tended to be fairly brief. Every few days, just like the proverbial bad penny or an irritating boil, Brent Spar would burst into prominence again with news of yet another confrontation between the environmentalists and the authorities, or another revelation about what was or wasn't in it, or what its disposal would or wouldn't do. But apart from the straight reporting of events and opinions, the affair was discussed endlessly. 'Disgusted of Tunbridge Wells' wrote to the newspapers by the sackload. Live conversations with the hijackers and Shell, or with government spokespersons, were broadcast almost daily on local and national radio and television networks. The Spar became a key subject of heated arguments on *Question Time*, while key figures in the story were mauled by Jeremy Paxman on *Newsnight*. It was the topic of editorial comment in innumerable periodicals ranging from specialist engineering journals through the heavyweight Sunday newspapers to *Private Eye*. It embroiled the highest levels of government, and there were heated exchanges about it in Parliament. As a result of its apparent support for Shell's disposal plans for the Spar, the UK Government was viciously attacked by politicians both at home and, more significantly, on the Continent. Finally, just before John Major's resignation, after Shell had given in to the pressures and had abandoned its original plan, the company's UK managing director apologised publicly to the Prime Minister for the embarrassment that the decision had caused. It was a big story by any standards. So what was it all about, and whatever happened to the battle-wearied Spar?

The basics of the story were simple, and fairly readily understood from the media coverage in the first few days of May despite lots of conflicting detail. The Brent Spar was an oil storage 'spar' buoy operated by Shell, or more correctly by Shell Exploration and Production (Expro) Ltd. Although it was operated by Shell Expro, and the media always reported it as owned by Shell, the Spar is actually owned jointly by Shell and Esso under a long-standing agreement between the two companies. Esso's financial interest was never widely publicised because the company wisely stayed totally out of the limelight – and therefore dissociated from the flak which all fell upon Shell.[1]

The Spar had been built in the 1970s and had been installed in 1976 in the Brent oilfield to the northeast of Shetland where, for a couple of years, it had stored oil from the two drilling rigs then in the field until it could be off-loaded into tankers for transport ashore. Although the Spar's main role was taken over in 1978 by a pipeline laid on the seabed from the Brent field to Sullom Voe in Shetland, it continued to operate in a subsidiary capacity until September 1991. At that time Shell decided that its operating value did not justify the expenditure that would be necessary to maintain it in a working condition and therefore took it out of commission. The company then had the problem of what to do with it.

The original hope was to get rid of it in the summer of 1993. But in view of the technical and legal complexities, not least because this was to be the first disposal of its kind for a North Sea installation, this hope turned out to be wildly optimistic. Following a series of studies between 1991 and 1994, Shell had decided that the best solution was to sink the Spar in the deep North Atlantic Ocean, and had applied to the UK Government for a licence to do so. In early 1995 the Government had announced its intention to grant permission for the deep-water disposal, but at the time of the Greenpeace occupation had not yet issued the final licence. Awaiting this licence, the Spar lay partly stripped of her contents and unmanned, still anchored to the seafloor at her old worksite.

In the meantime, Greenpeace had got wind of Shell's intention and decided that it was not acceptable. In true green activist fashion it had hatched its own daring plan to occupy the Spar, with the avowed intention of preventing Shell from doing anything with it until the autumn storms would, in any case, put paid to operations until the following spring. If the occupation could be achieved, it would also provide superb environmental publicity leading up to the fourth North Sea Conference of Environment Ministers from countries bordering the North Sea planned for early June 1995.

The occupation was achieved. With remarkably little difficulty, the Greenpeace vessel *Moby Dick* delivered the activists to the Spar from Lerwick in Shetland during the last weekend in April. On the Sunday afternoon, skilled climbers were first transported to the base of the buoy and then scaled the twenty-nine metres or so to the heli-deck above. Once they were there, and lifting gear could be installed, they were joined by other activists – and journalists taken along to report the proceedings to the world. The siege had begun.

Over the succeeding days and weeks anyone who cared to watch, read or listen was bombarded with a mass of information and misinformation. First, we learned more about just how big the Spar was, what it was made of and what various people thought it contained. We learned that the bit of the structure that was to become almost boringly familiar to us, the strange circular construction sitting on the top of what appeared to be a narrow stem disappearing into the sea, was metaphorically just the tip of the iceberg. This part contained mainly accommodation and spaces for the pumps, generators and various other pieces of equipment and machinery, while the main part of the Spar was hidden from view below the water. From the highly visible heli-deck at the top to the unseen ballast at the bottom the Spar was 137 metres tall, almost twice as high as Big Ben, with 109 metres, including six huge storage tanks, beneath the surface of the sea. Fully laden with oil it would weigh a massive 65,000 tonnes, though this figure is rather irrelevant since Shell would clearly not have left it full of such a potentially valuable cargo, filling the tanks instead mostly with seawater. There was much debate and argument later about just how much oil had been left in the tanks and pipework. But ignoring this, the weight of the Spar with its tanks empty was about 14,500 tonnes, about the same as a fairly big cross-channel ferry. Since it was built more or less like a ship, the Spar consisted largely of steel, making up about half its total weight, with most of the rest being iron ore ballast. The last seventy or so tonnes of the buoy's structural elements included a variety of other metals ranging from about thirty tonnes of aluminium to a few hundred grammes of mercury. Finally, as a result of its working life, the Spar had accumulated several other emotive components, including oily sludge, radioactive scale and dreaded PCBs. In summary, we learned that the Spar was big and dirty, but only about as big and dirty as a fairly average ship.

Next we learned of Shell's plans to get rid of the Spar by sinking it in the ocean. There was considerable confusion about just where it was to be sunk, with many erroneous references to the North Sea.[2] But the chosen disposal site was clearly identified as a spot some 150 nautical miles northwest of the Hebrides, in an area called the North Feni Ridge where the water depth is about 2,300 metres, almost a mile and a quarter. Greenpeace made it abundantly clear why it objected to this plan. First, it said, the Spar contained a variety of nasty materials, the effects of which in the deep ocean were quite unpredictable, but certainly bad. Second, if the Brent Spar deep-sea disposal went ahead it would create a precedent which would result in the offshore oil and gas industry dumping any amount of rubbish in the ocean. Third, Greenpeace claimed that onshore disposal was technically feasible, even relatively easy, and was not only environmentally acceptable but had some distinct advantages. It would admittedly cost more, but Shell should not be allowed simply to take the cheap option and ignore its environmental responsibilities. Fourth, the deep-sea disposal contravened all manner of promises and internationally agreed regulations against polluting the oceans. Finally, it was simply wrong to use the oceans as a rubbish tip.

Shell countered that Greenpeace was totally mistaken. The studies that the company had carefully conducted had compared a variety of disposal options for the Spar, including onshore dismantling. The deep-sea option had come out on top, they claimed. And it was not just because this option would be cheaper. It would, but not by much. But deep-sea disposal would be best on all counts, technical feasibility, safety to the workforce and, they claimed, environmental impact. Backed by the Government, Shell spokespersons also assured the public that, far from breaking any agreements or rules, they had played everything by the book both nationally and internationally. Moreover, there was no question of the Brent Spar creating a precedent for further deep-sea disposals. All such disposal plans would be treated on a case-by-case basis, that is, independently of anything that had happened previously. The battle lines were drawn.

At first, the media coverage was overwhelmingly supportive of Greenpeace and critical of Shell. So was the opinion of the public at large. It was a classic David and Goliath situation. How dare the vast industrial giant think that it could simply dump its unwanted rubbish in the pristine ocean instead of facing, and financing, its social and environmental responsibilities? Wasn't it great that those honourable eco-friendly men and women of Greenpeace, armed with only their rubber boats and unbounded personal courage and commitment, could stand up to the might of Shell backed by the UK Government? The pro-Greenpeace, anti-Shell feeling grew and grew as we were fed with images of the protesters being harassed by 'the authorities' and eventually being ejected from the Spar on 23 May after three weeks of occupation. This achieved, Shell could now continue with their plan and drag the Spar the 400 or so miles to the proposed dump site. Despite the widespread public protest throughout Europe, we saw Greenpeace vessels apparently vainly trying to stop the tow as the Spar inexorably approached the target. The battle was surely lost, but the war against industrial pollution would go on. But wait, it wasn't over after all. Shell was clearly on the defensive. Its public image, not to mention its sales of petrol at the pumps, had been severely damaged. Then, on 16 June, even more dramatic images hit the newsstands and the television screens. In spite of the presence of a defensive screen of tugs, support vessels with high-pressure water cannon, and even naval vessels, Greenpeace had managed to take possession of the Spar once more, this time by dropping two activists aboard by helicopter. Four days later they even had the neck to do it again! And now came the biggest shock of all. Within hours of the arrival of Greenpeace reinforcements, and despite categorical statements in the House of Commons to the contrary, even by the Prime Minister, Shell suddenly announced a change of heart. It was not going to dump the Spar in the ocean after all, at least, not immediately. Greenpeace and its supporters were understandably euphoric. Against all the odds, they had won.

After a couple of weeks of uncertainty, during which the British news media were preoccupied with the Major–Redwood battle for leadership of the Conservative party, Shell obtained approval from the Norwegians to moor the Spar in Erfjord, near Stavanger. The buoy could be towed to this location without having to be upended from its vertical position, a procedure which Shell had always maintained would be difficult and environmentally risky, and which was part of the case against its onshore disposal. But it would also allow the arguments about the contents of the Spar to be settled once and for all. In mid-July Shell commissioned Det Norske Veritas, an independent Norwegian maritime inspection and licensing authority, to carry out an unbiased audit.

Criticisms of Greenpeace's tactics and arguments had been growing for some time. Now that things were a little less frenetic, journalists, politicians, industrialists and even members of the public began to question the validity of 'single-issue pressure groups' in general and Greenpeace in particular. Questions were also asked about some of the so-called 'facts' that had surrounded the Brent Spar argument. In late August UK television executives even admitted that their coverage of the affair had been unbalanced, but claimed that this was because they had been misled by footage supplied by Greenpeace along with the environmental group's interpretation of it. In early September, about a month before the Det Norske Veritas (DNV) report was expected, Greenpeace admitted that its claim that the Spar contained much more oil residues than Shell's estimate was incorrect. Its excuse was that it had misinterpreted samples that had been taken by its activists during the occupation. When the DNV Report eventually appeared in October, it confirmed not only this error in the Greenpeace figures, but also demonstrated that in broad terms the quantities originally published by Shell for the make-up of the Spar, including the so-called 'contaminants', were essentially correct.

Although these developments prompted public comments about Greenpeace's 'loss of credibility' and 'betrayal of trust', the overwhelming opinion of the public, both in the UK and abroad, continued to be generally against the idea of deep-sea disposal. This was partly because of the power of Greenpeace's consistent and unequivocal ethical and environmental arguments, but also the fact that even 'expert' scientists appeared to disagree over some very fundamental questions.

Meanwhile the Government, who had originally issued the licence for the Spar's disposal in the deep waters of the UK's Exclusive Economic Zone, was still smarting from Shell's capitulation and the fact that its actions in issuing the licence were coming under considerable scrutiny. Accordingly, the Energy Minister, Tim Eggar, asked the Natural Environment Research Council to set up a committee of relevant scientific and engineering experts to examine the scientific and technical aspects of the various disposal options and their likely environmental consequences. When this committee produced its first report some five months later, it failed to support the Government position since it did not come down firmly in favour of the deep-sea disposal. Instead, the Report was widely, but wrongly, construed as supporting onshore disposal, the environmental consequences of which it had not, in fact, considered. But it also intimated that all was not well with the process by which the decision to issue a deep-sea disposal licence had been arrived at and the way it had been applied in the Brent Spar case. These criticisms were also largely ignored.

In the meantime, having apparently abandoned the deep-sea disposal option, Shell received more than two hundred unsolicited suggestions for alternative uses or disposal routes for the Spar from members of the public and from industrial and other organisations all over the world. In the light of this enormous response, coupled with the memory of the dramatic events of the early summer, Shell now began a long, complex and expensive exercise to win over 'hearts and minds', acknowledging that its previous PR efforts in relation to the Brent Spar had been less than successful.

But apart from winning over the public and the environmentalists, Shell still had a potentially major problem with the Government. Whatever it ultimately decided to do with the Spar, at least in the UK, it would need a government licence. The licence issued for the deep-sea disposal was still in force, and the Government had made it abundantly clear that, in order to obtain a licence for another option, Shell would have to demonstrate that the alternative was environmentally at least as good as, and preferably better than, the first. In other words, Shell would have to show that the Best Practicable Environmental Option (BPEO) used to obtain the original licence was not, in fact, the best! So the company announced publicly that it was starting the whole process again with no preconceived ideas, and that it would welcome input to the process from almost anyone. In particular, it solicited specific suggestions from industrial organisations who were experienced in dealing with engineering projects of this magnitude and would be capable of undertaking the disposal option ultimately chosen. Shell spokespersons and their press releases made repeated references to the fact that the deep-sea disposal, with its existing licence, was the baseline against which any alternative would be compared. Government spokespersons went further, emphasising that as far as they were concerned the deep-sea dumping option was still the preferred one. But it became increasingly clear that, having had its fingers burnt on the deep-sea option once, Shell was no longer likely to consider this to be a serious contender.

By July 1996, just about a year after the original capitulation, Shell had received almost thirty outline ideas for the re-use, disposal or dismantling of the Spar from nineteen potential contractors within

Europe and from as far away as Canada. Some of these suggestions were for the total dismantling and onshore disposal of the Spar, and one was for its burial in the shallow seafloor. But they also included ideas for its re-use either in its entirety or in part as an offshore wind-generated power collector, as an artificial reef, in harbour installations, for onshore oil storage, or for accommodation either at sea or on shore. From this long list of possible options, Shell Expro was to select about six of the potential contractors and their associated proposals to proceed to the next stage in which more detailed proposals would be prepared. After another selection process, those contractors whose proposals were thought to have the greatest potential to produce an acceptable alternative BPEO would be invited to submit fully detailed concepts. Finally, one, or perhaps a mixture of two or three of these detailed submissions would be submitted to the Government for approval and licensing. This stage was expected to be reached in the second half of 1997, more than two years after the change of heart, while the final disposal or re-use operation would not be completed for months, if not years, after this.

To assist Shell Expro in choosing between the various options at the different stages in this procedure, another element of the rethinking exercise, the Brent Spar Dialogue Process, was instigated. The objective was ostensibly for Shell to solicit the opinion of a much wider constituency in the decision-making process than it had previously done. But it was clearly also a public relations exercise. Indeed, to use Shell's own words, the hope was to 'engage not enrage'. In order to avoid later criticisms of 'fixing' the outcome, Shell engaged an independent organisation, the Environment Council, to run this Dialogue Process. To achieve this the Council arranged a series of one-day seminars to which delegates from a range of organisations were invited, the first held in London in November 1996 and subsequent ones in Copenhagen and Rotterdam early in 1997. It was made clear that the seminars were not decision-making events, but simply opinion and comment gathering exercises. The final decision on which option or combination of options for the disposal of the Spar would be chosen to go forward as the BPEO for licensing by the Government, would still be made by Shell Expro.

On 13 January 1997 Shell announced the shortlist of six contractors selected to continue with the disposal process and their ideas for the final fate of the Spar and the techniques to be used to achieve these ends. The ideas included total scrapping onshore, scrapping of the hull but re-using the topsides section as some sort of training centre, and re-use of the hull for a variety of purposes ranging from a dock gate, through quay extensions to fish farms and coastal protection modules. The suggested means to these ends included turning the Spar to the horizontal position using compressed air or raising it vertically either with compressed air or with a jacked cable lift, that is, essentially by crane. All the other ideas, including burial in shallow water, re-use at sea either for accommodation or energy generation, and towing across the Atlantic, had by now been rejected. The shortlisted contractors were asked to develop their ideas in more detail and to submit fully costed versions within a few weeks, but the deadline was later extended to early June when some of the contractors realised the scale of the problem. From these detailed proposals a final 'winner' would be selected to be submitted to the Government for licensing. In order to help Shell make this selection, but also to avoid accusations of self-interest, the services of DNV were again retained, this time as an independent assessor of the detailed proposals.

A second round of dialogue seminars was arranged in the autumn of 1997. Shell said that it expected to choose a 'preferred option' by the end of the year and to apply for a government disposal licence fairly early in 1998, after which they would be in a position to place a final contract or contracts for the Spar's disposal. Finally, in January 1998, Shell announced that the plan it was going for was the re-use of the bulk of the Spar in a quay extension for a ferry terminal in Norway. This, Shell claimed, was a unique solution for a unique problem and was not a precedent for the disposal of other offshore structures. Greenpeace, on the other hand, claimed ultimate victory and were confident that this was the death knell of any offshore disposal ideas. Either way, Shell was certain to get government approval for this new option. This would probably have been the case even without the change in UK government in 1997; with the new Labour administration it was even more certain in view of Labour's criticism of Shell's deep-sea disposal plan when the party was in opposition.

With no Greenpeace objections expected, within a year or two the bulk of the Brent Spar will have been cut into a few large pieces hidden under a concrete jetty, while the rest will have been recycled into the proverbial ploughshares. A few months further on, hardly anyone other than those directly involved will even remember the Spar.

So why write a book about it? After all, we now know what the final fate of the Spar will be. Why rake over the coals yet again? Well, for lots of reasons. First because it is a curiously intriguing story. It may not have much sex and violence, apart from a few shoot-outs at Shell petrol stations in Germany, but it has most of the other ingredients of a real-life whodunit. Good opposed to evil; the common person against bureaucracy; morals versus money and expediency; aesthetics against cold unfeeling science; and above all David triumphing over Goliath. How often do the direct actions of the person in the street bring a major industrial organisation to its knees?

This is all good stuff, and although bits of it have been told many times, some important parts have not. They raise questions that could do with a further airing. If the deep-sea disposal plan was such a bad idea, how did Shell choose it in the first place, how did it think it could get away with it, and how did it convince the Government to give it a licence to do it? And in view of the opposition of other European governments, by what right did the UK Government think it could issue such a licence? Who owns the seas, and what about all these international regulations that Greenpeace maintained the Government was bound by? What about the business of precedent? How many offshore installations would the industry like to dump in the sea, and if the Brent Spar had been dumped successfully would the oil industry have taken this as *carte blanche* to do the same with the rest?

But there is also another layer of the Brent Spar cake that needs looking at more closely. The so-called 'facts'. How could the two sides have such differing views on what was in the Spar, and, more importantly, on what it would do if it were dumped on the bottom of the deep sea? Both sides used 'science' to back up their arguments, and during the ensuing debate various scientists not involved in the original decision jumped in on one side or the other with their own two-pennyworths. What were the scientific 'facts', and how much reliance should we put on them? What is the non-scientist to make of a situation where scientists seem to disagree amongst themselves? More generally, what is the role of science and scientists in these situations? Is science more or less important than other factors like costs, politics or ethics?

Most important of all, and embracing all of these reasons, is the light the Brent Spar affair throws on the process by which these difficult decisions are reached. Assuming that the ultimate solution for the Spar is the 'best' one, was the process of getting there an acceptable or reasonable one? Few would these days argue against the existence of, or indeed the need for, pressure groups of various sorts, including environmental ones. But is the confrontational approach and the resulting public pressure that typified the Brent Spar affair a good way to reach the best solution? Do laudable ends, even if these can be agreed, justify any means? In this case, it was not as though the two sides had totally different agendas, as is certainly the case in many of the other causes taken up by Greenpeace such as whaling or sealing. On the face of it, both Shell and the UK Government on the one hand, and Greenpeace on the other, were anxious to find the 'best' solution to the problem of getting rid of redundant offshore structures. Whatever you may think of the Greenpeace approach, it must be given credit for opening up the debate and for revealing some very murky areas. Surely these areas should have been scrutinised and clarified long before the environmental activists climbed up the ladders onto the Spar. Why were they not?

So far, the arguments here have applied specifically to the Brent Spar, or at least to the present and future problems of the North Sea offshore oil and gas industry. But there is a much broader context too. Whether you are a Euro-sceptic, a Euro-enthusiast, or something in between, you can't ignore the fact that the UK is geographically close to continental Europe, whatever you may think of any political or social 'closeness'. As environmental issues are so often trans-boundary, the UK's environmental problems are also Europe's, and vice-versa. Depending on one's viewpoint, one may accordingly feel that the influence on events in the UK of the actions, for example, of the German public, or the statements of ministers of a range of continental countries, were either an unwelcome infringement of our sovereignty by outsiders, or an expression of European unity. But even if you think this external interference was a good thing, you will surely acknowledge that the way it came about was a little chaotic. Even in the absence of the broader global argument, the European dimension makes the Brent Spar story worth looking at again for lessons that might help to avoid the same or similar consequences next time. Again, there has got to be a better system.

Those are the reasons for writing the book. Now, what can you expect to find in it? First, a description of what the Spar was. Then a short, and exceedingly simplified account of the whole offshore decommissioning area. There are hundreds of books and conference proceedings devoted to the complex world of decommissioning offshore structures, and it is not our intention that this book should explore the area in any great detail. However, some background information on the subject is required to understand some of what went on during the Spar incident. Then some of the scientific aspects of the argument are discussed, along with the whole thorny area of 'precedent'. We then present a blow-by-blow account of what happened. You have already read a very brief account of events from the initial occupation to the present time. The final chapters deal with the aftermath of the event and what it has meant for Shell in particular, and industry in general. Finally, there is a personal viewpoint on whether we thought that the 'siege of the Spar' was a 'good or bad thing'.

As a final comment, it seems that although the fate of the Spar is now, to all intents and purposes, sealed, the whole sticky issue of decommissioning oil platforms has been ignited again by the new Labour Government. Michael Meacher, the Environment Minister, at a press conference in May 1998 on the eve of an Ospar Commission meeting on the protection of the marine environment, insisted that offshore dumping of installations would continue to be an option for up to sixty North Sea installations. Accusations that Mr Meacher was reneging on his earlier pledge, made the previous September, to 'presume against' sea dumping have been levelled at the Minister. This apparent change of stance is made more interesting by an earlier quote from Mr Meacher, then as Shadow Environment Minister, that there would be 'no more Brent Spars under Labour'.

As this book went to press, the results of the Ospar Commission meeting in July 1998 were eagerly awaited.*

---

*   In the event, and to the dismay of the UK Offshore Operators Association, the OSPAR meeting agreed that all North Sea structures weighing less than 10,000 tonnes must be totally removed. At least the topsides of all larger structures would also have to be removed and, wherever practicable the 'footings' also. See also note 1 to chapter 2.

Chapter 1

What was the Brent Spar?

THE NORTH SEA OIL INDUSTRY – A BRIEF REVIEW

Before discussing exactly what the Brent Spar was, and more emphatically what it was not, it would be prudent to sketch out the background to the development of North Sea oil and gas, to put the Spar's existence and *raison d'être* into context.

During the late 1960s and throughout the 1970s, the North Sea province became increasingly attractive to oil and gas exploration companies in their quest for new and politically stable offshore reserves. Until this time, interest in offshore exploration in the North Sea developed sluggishly as a consequence of the vast quantities of low-cost oil available from the Middle East, North Africa and other parts of the world. These areas seemed to offer a stable long-term supply of oil to satisfy the world's rapidly increasing demand for fossil fuels.

However, by the late 1960s political and economic activities in the oil world were beginning to concern oil companies, convincing them to reassess the potential importance of western offshore oil deposits. The Organisation of Petroleum Exporting Countries (OPEC), formed in 1960 to safeguard the interests of the oil producing nations, comprised thirteen of the leading oil exporting countries by the early 1970s. The combined output of OPEC members represented 90 per cent of the oil traded internationally at that time. From 1971, the breakdown of the traditional oil system organised by the multinational oil companies and the gradual takeover of power by OPEC, resulted in the two world oil crises that saw oil prices escalate.

The first crisis began in 1973, when events in the Middle East, combined with changes in the international oil community, necessitated a complete overhaul of the security of important oil supplies. In 1973, oil constituted nearly half of the world's annual use of 'industrial' energy forms (as opposed to 'traditional' energy forms such as crop wastes, fuelwood and dung) and the sudden rise in price doubled the real price of oil on the world market. The crisis had a greater effect than expected on the energy market as a whole, as there had previously been decades of constant or declining monetary costs of energy. These issues added incentive to the oil companies' quest for new North Sea reserves, and substantial quantities were indeed discovered and exploited throughout the 1970s, albeit in very challenging and tempestuous conditions. Within a decade the North Sea oil industry was constructing production installations that ten years previously had been considered impossibly ambitious. By the late 1970s, London and Aberdeen were becoming oil centres of excellence, and companies such as Shell, Exxon and BP were heavily invested in the North Sea sector.

In 1971, Shell/Esso discovered the Brent field in water depths of about 450ft, whilst drilling in the most northerly offshore well in the world at that date. With this significant find, and the discovery of the Forties field off Aberdeen in 1970, the status of the North Sea as a major oil and gas province was firmly established. These two biggest fields were estimated to provide enough oil to meet total UK demand for nearly five years (Taylor and Turnbull 1992). Although it was not technically feasible to recover Forties and Brent oil at a rate fast enough to cover more than about half of the daily UK demand, other established commercial finds during the first few years of northern North Sea exploration offered the prospect of UK self-sufficiency in oil. In fact, by 1980 the UK was quantitatively self-sufficient in oil, and by 1981 UK oil production exceeded consumption for the first time.

The North Sea now ranks as a mature oil province with a network of production facilities in place. It has been one of the most prolific oil basins outside of the Middle East. More oil remains to be found in the North Sea, although it is doubtful whether there are more large fields such as Forties and Brent to be discovered. Finds are more likely to be made in the smaller, marginal fields, and the search for these hard-to-find fields is well under way. It is thought that future oil finds are likely to be in the 50 to 100 million barrel range (Brent and Forties are both 2×10^9 barrel fields) but the state of knowledge is such that larger oilfield finds cannot be categorically ruled out.

Over the thirty years of North Sea operations, operating costs have varied from approximately US$10 to $16 per barrel of oil, compared to about $2 a barrel for certain types of Middle Eastern oil. Recent history and events in the world energy market over the last twenty-five years, however, have borne out the tremendous economic and technical risks that the oil industry undertook with the development of North Sea reserves. In the first year of production, the North Sea was averaging less than 250,000 barrels a day. By 1976, it was producing 1 million barrels a day and by 1985 it was producing its peak output of 2.6 million barrels per day.

THE BRENT SPAR

Once the presence of oil in sufficient quantity has been discovered, there are a number of onshore and offshore activities that need to be undertaken to allow exploitation and supply of that reserve. As soon as the field is declared commercially viable, the provision of concrete and steel production platforms, storage facilities for the oil, terminal facilities for the landing and primary processing of oil and gas and onward transmission have to be considered and implemented. The first offshore platform for the exploitation of hydrocarbons, installed out of sight of land, was developed in the Gulf of Mexico during the 1940s; since that time hundreds of offshore installations have been commissioned to facilitate the exploitation of oil and gas reserves.

The North Sea, in particular, introduced some interesting problems as far as offshore operations were concerned. The area is infamous for unpredictable and particularly harsh weather conditions. Violent storms can blow up within an hour and can last for days. Even in the shallower waters of the southern North Sea, the industry realised the need to be prepared to ride out storms that could create freak conditions of waves rising to 30m and wind gusts up to 150 knots – a phenomenon known as the 'hundred year storm'. Northwards, the conditions were even more unpredictable, and these weather uncertainties added to the already significant risk of placing rigs into an area with no established oil or gas prospects at the time.

Whatever the perceived difficulties of installing offshore equipment, offshore operators lost no time in looking into ways through which the oil and gas reserves could be developed. The complexities of operating offshore at the deepest locations in the North Sea involved investment in some of the most expensive projects the industry had yet undertaken. The development of the Brent field complex and its associated share of pipelines and terminals cost over £3 billion, comparable to the cost of a moon expedition at the time (Taylor and Turnbull 1992).

The two main types of structure suitable for offshore developments are steel platforms and concrete platforms. In the case of steel platforms, the main structure (or 'jacket') is built on shore, towed out to the field in a horizontal position, up-ended on location and secured by piles driven into the seabed. The concrete or gravity-type platforms are kept in position by their own weight and settle firmly into the seabed and, since they need a large base, have ample room for storing the recovered oil. As no piling is required they can be installed much more quickly than steel structures. The concrete structure, with its inbuilt storage, may then seem to have a distinct advantage over the steel structure. However, the feasibility of combining storage with an offshore loading facility in a steel-based installation, was proved by the novel approach taken by Shell in the use of the Spar floating oil storage and loading facility in the Brent field.

When the development of the Brent field was first being considered, Shell was looking for a relatively deep-water floating installation which would have sufficient storage to sustain output from the initial Alpha wells, which had no storage capacity. The intended capacity would also eliminate potential shutdowns caused by delivery interruptions each time a tanker left its mooring when full, or as a result of deteriorating weather. The storage capacity chosen for the installation was to be the equivalent of three days' production, or 41,000 tonnes of oil. A 'spar' floating buoy was conceived and designed to serve this purpose (North Sea Fields 1987). It was installed in the Brent field in the summer of 1976 (figure 1). Interestingly, an alternative and more conventional solution, favoured by Shell's partner Esso, was to use a more-or-less customary tanker attached permanently to a single-buoy mooring (Howarth 1997). Esso's solution was considered to be much the cheaper, both in the short and long term, partly because a conventional tanker could be used for other purposes whereas a spar buoy was assumed, correctly as it turned out, to have no salvage value. Nevertheless, Shell's more expensive (total cost was in the region of £24 million) and revolutionary idea was ultimately selected mainly because it would result in less pollution from oil spills. This at a time when the government pollution regulations were extremely unclear and certainly did not oblige the industry to take this sort of decision.

The 'Brent Spar' was, and still is, frequently referred to as an 'oil rig'. To most people an oil rig is a structure used to drill into the earth to reach and extract oil. This general term covers a multiplicity of structures and an enormous range of size and complexity. It also covers the vast majority of the structures in the North Sea offshore oil and gas fields. But the Brent Spar was emphatically not one of these. It had nothing to do with the extraction of oil from beneath the sea, but only with storing it.

The Spar is therefore basically an enormous oil tank, or rather six separate tanks clustered around a central core rather like the segments of a divided orange. Technically, it was a spar buoy designed to float vertically, like a huge fishing float. It was originally built as a specific solution to a particular problem. So the Spar was extremely unusual, if not unique, both in the sense that there is no other structure exactly or even reasonably similar to it, but also because there are probably no more than two or three others in the North Sea that are based even vaguely on the same principles.

Under normal circumstances the bulk of the Spar, including the whole of the oil storage facility, was submerged and therefore invisible (figure 2). But in its entirety it resembled a vast bottle with a rather large cap, with the surface of the sea about halfway up its relatively narrow 'neck' (see figure 3).

With its tanks empty, the Spar weighed 14,500 tonnes, about the same as 2,000 double-decker buses or a modern, large cross-channel ferry. The *Stena Fantasia*, for example, weighs a little more than the Spar at 15,710 tonnes. Again, comparable with the *Stena Fantasia* at 163m long, the Spar's total length was 137m, of which only the upper 28m protruded above the water surface in its working condition. This part, with which we all became familiar as a result of the endless media coverage, consisted of a 26m-diameter superstructure made up of a machinery deck, a diving deck and an accommodation deck surmounted by a turntable with a helicopter landing pad, cargo crane, loading boom, mooring hawser and hydraulic power pack. Beneath the machinery deck the Spar narrowed to a column (the bottle's neck in our analogy) 17m in diameter and 32m high, half above the water and half below. Like the superstructure, the column was divided into a series of decks for specific purposes including metering the flow of oil in and out of the Spar, anchoring procedures, equipment storage and, immediately above the water level, housing for the main pumps.

Below this relatively narrow neck, and beginning therefore 16m beneath the sea surface, the main body of the Spar was a 93m-long cylinder, 29m in diameter. The bulk of the body was made up of the oil storage tanks weighted with ballast at the bottom, and buoyancy tanks at the top.

The ballast consisted of 6,800 tonnes of haematite, or iron ore, held together in concrete and housed in twenty-four separate compartments occupying the lower 3.5m of the Spar. At the other end of the main body, the upper 13.8m housed two series of six air-filled buoyancy tanks, the inner series within an extension of the column into the main body.

Finally, between the ballast and the buoyancy, the main body was made up of six oil storage tanks arranged radially around the central shaft, a circular space running up through the entire structure. The oil arrived from the production rigs through pipes, or risers, in the central shaft and was then pumped through the Spar's own pipework into the various tanks. The structure of this lower part of the Spar was very similar to that of a ship, with a thin outer skin of 20mm-thick steel plate stiffened by ribs and bulkheads. Each tank was 75.95m high and had a capacity of 50,000 barrels (7,900m^3), giving the Spar a total capacity of 300,000 barrels (47,400m^3). If the Spar, tanks entirely filled, could have been hung from a giant spring balance it would have weighed in at about 65,500 tonnes. Even given a big enough balance, the Spar could never be weighed in this way because the thin walls of its tanks could not contain the contents without bursting unless supported by the pressure of the surrounding water. In any case, although this figure appeared in many media accounts at the height of the Greenpeace occupation, it is entirely misleading since the contents of the tanks, mainly seawater in its decommissioned state, are irrelevant to its size as far as disposal is concerned.

But the total oil capacity of the Spar was, in any case, never used for its intended purpose because two of the tanks were damaged after less than one year of operation. In fact the Spar was damaged prior to beginning its working life. Before it was installed in the Brent field, the Spar had to be gradually up-ended to the vertical by slowly ballasting the storage tanks; this nine-day operation took place in a Norwegian fjord. During the operation, the Spar buoy was found to be overstressed by the pressures it experienced. To compound matters, in January 1977, just one month after it commenced operations, two of the storage tanks were ruptured during a build-up of differential pressures which were in excess of the design limits. The repairs only went as far as maintaining the structural integrity of the buoy, i.e. they were not made watertight. As a result, these tanks were left open to the sea and were used only to balance the pressure in the other tanks as they were filled and emptied of oil, and for the temporary storage of the water separated from the crude oil as it was received from the drilling rigs.

To fill and empty the storage tanks, and to transfer oil and water from one tank to another, the Spar had a complex network of piping connected to the pumping system. In addition, each of the storage tanks was fitted with a 6in-diameter vent pipe extending to A deck some 32m above them. These vent pipes played an important role as the public saga developed because they were used by both Greenpeace and Det Norske Veritas to provide access to the storage tanks to obtain estimates of the amounts of oil remaining in them.

During its working life the Spar was anchored firmly in position by a six-leg 'catenary' mooring system. Catenary simply means the curve adopted by a chain or wire suspended from two separate points. The cables of a suspension bridge, for example, adopt a 'catenary'.

Having been built rather like a ship, the bulk of the weight of the Spar, apart from the 6,800 tonnes of ballast, was made up of about 6,700 tonnes of structural steel. The remaining 1,000 tonnes or so consisted of electrical wiring, the sacrificial anodes, of which more than a thousand were bolted onto various parts of the underwater sections of the Spar to reduce corrosion, several tonnes of paint, and the equipment that had been left aboard. When the Spar was taken out of service, much of the equipment with which it had been fitted during its working life was removed, particularly those items that could be removed relatively easily or had some significant re-use value. But a good deal was left in place, including the main pumps and some of the large items of fixed equipment such as transformers, fire pumps, emergency generator, transformer and gas turbines. At the time of the Greenpeace occupation it also contained a variety of materials that it had picked up during its working life. Some of these, such as the oil residues and sludge, and the radioactive scale lining the tanks and pipework, were measured in tonnes, and we will come to them when we look at the Spar's detailed make-up as subsequently determined in an independent survey. Others revealed by the on-the-spot television pictures included the debris left by the last workforce to occupy it, ranging from discarded bedding and books to cooking utensils and tins of baked beans. Although these were pretty irrelevant in a global sense, and while Shell may have intended to clean it up before its final journey to the bottom of the ocean, the impression was that it was a sort of *Marie Céléste* rubbish tip.

Before the Greenpeace occupation of the Spar, and particularly as part of the licensing documentation, Shell released detailed inventories of what materials it claimed were in the buoy. In addition to the major structural steel and ballast components, Shell listed various additional materials, including 'contaminants', ranging from about 50 tonnes of oil residues, almost 30 tonnes of aluminium and 13.5 tonnes of copper, down to 0.3kg each of arsenic and mercury and about 19ml of PCBs. Greenpeace used these same figures in their own publicity releases, but questioned their accuracy since many of the figures were, on Shell's admission, based on estimates rather than accurate measurements. Greenpeace therefore argued that many of the more toxic components, particularly, had been 'grossly underestimated'. In addition, Greenpeace claimed to have evidence that unspecified materials had been secreted in the Spar, while from its own sampling of the tanks during the occupation it maintained that these contained some 5,500 tonnes of oil and sludge rather than the 50 tonnes suggested by the Shell figures. If the planned deep-sea disposal of the Spar had gone ahead, of course, these disagreements would never have been resolved. As things turned out, however, they could be.

In July 1995, following the decision to postpone the disposal of the Spar and to tow it instead to Erfjord in Norway, Shell commissioned the Norwegian organisation Det Norske Veritas Industry AS (DNV) to undertake an independent inventory of the structural components and contents of the Spar, particularly in relation to the conflicting views that had been expressed.

As the Norwegian equivalent of Lloyds, and with its own expert and highly experienced staff, DNV was well qualified to carry out such an inventory. DNV began the inventory investigation at the beginning of August 1995, and it was completed when the results were presented to Shell and the UK authorities on 16 October 1995 and made public at a press conference two days later (see DNV

Reports 1995). To ensure the independence of the work, none of the results were disclosed to anyone, including Shell, prior to the presentation. Responsibility for the safety of the Spar and of all personnel aboard remained with Shell throughout the work and a Shell safety representative had to be on board every time the project team visited the structure. But the representatives took no part in any sampling or inspection procedures, nor were they included in any discussions of the findings.

DNV had three main tasks. First, to check the figures given by Shell for what the Spar was actually made of, and particularly the non-structural metals in the sacrificial anodes, batteries, electrics, paint and so on; second, to check on the 'contaminants' such as oil residues and radioactive scale that had accumulated in the Spar during the course of its working life; and third, to ascertain whether any additional unlisted 'nasties' had somehow got into the Spar.

The DNV Report broadly confirmed Shell's figures for the Spar's make-up as far as metals were concerned, with the Shell estimates generally falling within the ranges determined by DNV. There were a few minor exceptions that no-one seemed to think were particularly important. For example, the DNV estimates for amounts of aluminium in the Spar were rather higher than Shell's, while those for zinc were significantly lower. This discrepancy seems to have arisen because while Shell had assumed that the sacrificial anodes were zinc-based, DNV assumed that they were aluminium-based as specified on the maker's original drawings.

The single significant contrast with the Shell figures in this part of the DNV Report was for PCBs, or polychlorinated biphenols. These chemicals are among the most damaging pollutants known, being readily bio-accumulated, that is becoming concentrated in animal tissues, particularly in fats, and having very serious deleterious effects on a variety of physiological functions. Consequently, the amount of PCBs in the Spar was an important factor in the Greenpeace argument against the deep-sea disposal option.

Because of their toxicity, the use of PCBs is now banned, but until the mid-1970s they were used in a variety of situations, particularly in electrical fittings. Shell's original figure of 19ml was not measured, but was based on the assumption that traces of PCBs remained in the Spar's transformers. When the Spar was built, PCBs were still being used extensively, and the transformers initially installed were filled with PCB-containing fluids. Some years later these fluids were removed and taken ashore for disposal by high-temperature incineration, the only known effective way of destroying PCBs. The transformers were then flushed out twice and refilled with non-PCB-containing fluids. But it was assumed that despite this treatment a small amount of PCBs would remain, hence the rather curious figure of 19ml, presumably based on the surface area of the insides of the transformers and some estimate of the thickness of the layer of PCBs that would remain adhering to these surfaces.

But PCBs were also widely used in other electrical appliances and particularly, for instance, in the capacitors of light fittings. Because it was short of time in producing the original report against Shell's deadline, DNV did not have the opportunity to examine the light fittings on the Brent Spar. But since PCBs were still in widespread use when the Spar was built, it assumed, probably correctly, that the fittings originally installed would have contained PCBs in their capacitors. DNV accordingly did a simple calculation and came up with a figure of 6.8–8kg of PCBs which was immediately seized upon by Greenpeace and its supporters. Shell denied these figures, claiming that the PCB-containing fittings had all been replaced later in the Spar's life. But despite DNV's corroboration of the other figures for the Spar's make up released by Shell, no-one would take their word for the PCB contents, the most charitable interpretation being that Shell had been somewhat cavalier in not undertaking the relatively simple task of removing the offending fittings in readiness for the proposed dumping. Accordingly, Shell asked DNV to carry out a supplementary survey, this time concentrating specifically on the possible locations of PCBs in the Spar. This survey was carried out on 20 November and was the subject of a fourth report, 95–3583/2. This report described how DNV examined all the remaining light fittings (and two tumble dryers!) and confirmed that they were all, as Shell had claimed, purchased and installed after PCBs were no longer used in their manufacture.

DNV's second task, that of calculating how much the Spar had become 'contaminated' during its working life, particularly with radioactive scale and oil residues, was a little more complicated. The presence of the dreaded radioactivity in the Spar was always openly admitted by Shell. But despite lots of accounts of where it came from and assurances that it was not at all dangerous, there was a widespread suspicion among the public that Shell's activities had either contaminated the Spar with extraneous radioactivity which would accompany it to the seabed, or that they were using the structure and its proposed fate to get rid of something they did not want and which would be more difficult to dispose of in some other way. Neither of these scenarios turned out to be correct.

The accumulation of so-called Low Specific Activity or LSA scale on the insides of tanks and in pipework is a well known phenomenon in the oil industry. The radioactivity in the scale arises from naturally occurring radionuclides that are originally in the rock strata in which the oil is found. The water associated with the oil reserves, called formation water, tends to dissolve the radioactive salts from the rocks. When this water is brought up with the oil, and particularly if it mixes with seawater, it will deposit a scale, mostly of barium sulphate, on the insides of tanks and pipes through which it passes. The radioactive salts accumulate in this scale, making the scale itself mildly radioactive. Some of the salts also accumulate in the oily sludge at the bottom of storage tanks so that this also becomes slightly radioactive.

The scale from North Sea operations forms a very hard, ceramic-like layer which is resistant to dissolution by strong acids and bases, and is even quite difficult to remove physically. When it must be removed, one of the preferred methods is the use of high-pressure water jets, since if it is mobilised into airborne particles by, for example, hammering, cutting or grinding, there is a risk that it might be inhaled and deposited in the lungs. Although it is only mildly radioactive, its removal in this way poses a small, but significant, health hazard to the workforce,

It is also difficult to estimate the volume or weight of LSA in a particular situation. Shell more or less guessed that the Brent Spar might contain about 30 tonnes of scale, assuming a worst-case scenario in which all 1,000 metres of the relevant 25cm-diameter piping was coated with a uniform layer of scale 10mm thick. DNV and its associates thought this was an overestimate by a factor of about three, because much less of the pipework appeared to be scaled than Shell had feared. As a result, Shell estimated the total radioactivity in the scale at around 6.9 giga-becquerels (a becquerel being the standard unit for measuring such activity) whereas the DNV report suggested a maximum scale activity of about 2.2 giga-becquerels. On the other hand Shell had underestimated the total activity in the sludges in the bottom of the tanks at 5.1 giga-becquerels compared with the DNV estimate of up to 13.8 giga-becquerels. Nevertheless, the estimates of the Spar's total activity from all sources (12 giga-becquerels by Shell and up to 13.8 giga-becquerels by DNV) were very similar. Moreover, since the type of activity emitted by the scale and sludge is very like that produced by granite rock, the much repeated point that exposure to the activity in the Spar was equivalent to that from a cluster of buildings in Aberdeen could be made with some justification.

A more difficult problem was to assess the amount of hydro-carbon or oil remains that had accumulated and been left in the Spar. As part of the decommissioning process, Shell had drained oil from the piping system and also removed most of it from the storage tanks. Nevertheless, it estimated that about 53 tonnes remained, mostly in the water, sludge and waxes in the tanks. According to DNV this was an underestimate, mainly because Shell had ignored the oil left in dead spaces in the pipes. The true figure, said DNV, was up to about 130 tonnes, but most likely in the region of 74–103 tonnes. But this discrepancy paled into insignificance beside the figure for the oil that Greenpeace claimed had been left in the Spar.

This claim was based on samples that Greenpeace activists took themselves during the occupation. Since they did not go aboard the Spar prepared for this task, they had to make do with what was available, apparently an empty food jar on the end of a piece of string! With this sounding device they attempted to find out how much oil was left in the tanks. When the resulting samples had been analysed at the Greenpeace Research Laboratories, Greenpeace announced in mid-June, shortly before Shell's change of heart, that it had found more than 5,000 tonnes of oil in the tanks. At the time, although Shell disputed the Greenpeace claim, the details of how the Spar's occupants had taken their samples was not subjected to close scrutiny. However, during the subsequent survey DNV were naturally interested in these details. At a meeting between DNV and Greenpeace in early August the impression was given that the samples had been obtained from M deck, that is immediately above the tanks and 29m below sea level. From DNV's knowledge of the Spar's construction, and particularly its pipework, it doubted that this was possible but did not express its reservations to Greenpeace. Instead, DNV subsequently asked Greenpeace for confirmation that all its oil and water samples had been taken from M deck. Consequently, Greenpeace wrote to DNV to confirm that its initial explanation about how the samples were taken was incorrect, and that they had actually been taken from the 6 inch-diameter vent pipes which lead from the tops of the tanks to open on A deck some 28m *above* sea level. So the Greenpeace samples had detected oil floating on the top of water only in these vent pipes and not in the tanks themselves which were, in fact, mainly filled with seawater. In late September 1995, some three weeks before the DNV Report was to be published, Greenpeace openly admitted this error and retracted its figure for the oil content of the Spar.

DNV's third task was to find out if the planned disposal of the Spar had been used to get rid of other unwanted rubbish that had nothing directly to do with the buoy or its working life. This had been prompted largely by the 'concrete egg' story, one of the most intriguing and possibly influential red herrings of the Brent Spar affair.

According to this story, which received considerable media coverage in mid-June 1995 just before Shell's capitulation, some unspecified, but presumably highly hazardous material had been secreted somewhere in the Spar but had not been mentioned in Shell's inventory. The background to the story, and the ultimate outcome, is well told in DNV's topside inspection report from which the following is largely taken.

'In the course of a television programme Greenpeace produced an affidavit which had been signed on 13 June 1995 by a Mr John Lennie and duly notarised by a solicitor and notary public, John M. Johnston'. (Apparently, Greenpeace had been approached by Mr Lennie and the affidavit was produced for him, with his agreement, by Paul Horsman of Greenpeace.)

According to this affidavit, and subsequent interviews between Mr Lennie and DNV, between 1978 and 1987 Mr Lennie had been employed by Deutag Drilling as a roughneck and had worked on a number of different platforms including Brent Alpha and Brent Bravo. In November 1980 Mr Lennie and two American colleagues were sent to the Brent Spar because of a 'difficult and dangerous situation' that had arisen. When they arrived they found a Red Alert in operation, resulting in the evacuation of the Spar's personnel. Mr Lennie claimed that they were told that three tanks of chemicals, each of about 500 gallons, had been delivered to the Spar but had been wrongly labelled. They had been emptied into some unspecified tank and had 'caused a reaction which was the reason for the Red Alert'. Mr Lennie further claimed that he 'could feel the heat and a bad smell of rotten eggs'. He claimed that he was instructed to connect up some unspecified pipework so that cement under high pressure could be pumped into the problem tank to surround and seal off the reacting chemicals and so produce the 'concrete eggs'.

During their attempt to investigate this affair, DNV tried to arrange a visit by Mr Lennie to the Brent Spar, then already moored in Erfjord, in order to try to identify the area of the Spar where the 'eggs' were supposed to be. The first attempt, in early September 1995, was aborted when Mr Lennie became ill during a flight between Oslo and Stavanger and decided that he was not fit enough to take the helicopter flight to the Spar. Accordingly, a second visit was arranged, starting from Aberdeen, on 19 September.

On this occasion Mr Lennie met the DNV team at an Aberdeen hotel during the previous evening and, over dinner, confirmed his earlier account of the affair, but declined the offer of reading through his recorded statements until after his visit to the Spar. But the next morning Mr Lennie failed to appear in the hotel reception area as arranged. When traced to his room he was found preparing to leave and announced that he would be unable to visit the Spar because he had received a message telling him that his cousin had had an accident. The DNV team left and subsequently spoke on the telephone with Mr Lennie's wife, who told them that it was, in fact, her brother (Mr Lennie's brother-in-law) who was ill and had been hospitalised following a nervous breakdown.

Having failed to get Mr Lennie onto the Spar to identify more accurately the locality of the supposed 'eggs', DNV personnel examined all the tanks and pipes that they thought could possibly have been implicated, except the inner buoyancy tanks which were inaccessible for safety reasons. No sign of any cement could be found except around the anchor chain leading to the chain lockers, and this had clearly not been pumped into place under pressure.

DNV then requested access to the official log books that every offshore installation is obliged to maintain at all times and which have to be deposited with the Department of Energy at six-monthly intervals. 'For the period of interest', says the DNV Report,

> The HSE [Health and Safety Executive] in Aberdeen was only able to produce the log books for the period December 1981 to December 1982. In these books DNV found no evidence of incidents which could explain Mr Lennie's version of events. Subsequent correspondence with the HSE indicates that log books prior to 1982 may have been destroyed when responsibility for keeping official log books for offshore structures

passed from the DOE [Department of the Environment] to the HSE.

From their investigations, including the interviews and their examination of the Spar, DNV came to the following conclusion:

> As a result of the investigation DNV has not been able to find any evidence to support the affidavit claim for the existence of the 'concrete eggs'.

By this time, of course, arguments about the existence of concrete eggs, like those about the amounts of PCBs and oil in the Spar, were largely academic. However, in retrospect it is possible to say that the figures for the make-up and content of the Spar provided by Shell were in all essential details correct.

WORKING LIFE AND EARLY RETIREMENT

As discussed earlier, two of the Spar's tanks were damaged during installation. They were strapped up to maintain the Spar's overall integrity, but no attempt was made to reseal them. They were therefore used subsequently only as settling tanks to separate water from the oil stored in the intact tanks, and not to store oil themselves. When oil is first extracted from the oil-bearing strata in offshore drilling operations it usually contains a certain amount of water. This water, called formation or produced water, is separated from the oil and discharged back into the sea. Sometimes this separation is undertaken at the drill rig. But there was no water-separating equipment on the Brent Alpha rig and the water in its oil was therefore separated on the Spar. Once the oil and water mixture was in the undamaged storage tanks, the water sank to the bottom and was then transferred to the damaged tanks. Here, it would mix with 'old' produced water from previous separations and with 'normal' seawater entering though the cracks in the outer skin. Eventually, the produced water and normal water mixture would escape into the surrounding sea. But before this happened, the mixture would stay within the damaged tanks for a further separation of any oil carried over with the produced water, so that the final liquid released had a residual hydrocarbon content of less than 40mg/l (DNV Management Summary Report, 15).

In the early days of its working life, the Spar was the sole route by which oil was exported from the Brent field. But in 1978 the Brent System Pipeline was commissioned. This system consisted of a 24in-diameter pipe laid on the seabed to carry the Brent oil to the Sullom Voe terminal in Shetland. Thereafter, the Spar was used only as an alternative storage and offloading facility in case of problems with the pipeline. So within only a couple of years of its installation, its main function had more or less disappeared. Nevertheless, the Spar was employed profitably in a supporting role for rather more than ten years. But by the late 1980s it was in need of considerable attention, and maintenance costs increased significantly between 1987 and 1990. In 1991 an internal review concluded that if its working life was to be extended it would need £90 million spending on it and it would be out of commission for 2–3 years for the refurbishment. In view of the Spar's age and the existence of the pipeline system it was decided that the buoy should cease operations. Accordingly, in September 1991 it was taken out of commission after a total of fifteen years' service. The working storage tanks were emptied of crude oil and filled with seawater. The process pipework was flushed through with seawater and the storage tank oil/water interface emulsion and slops were pumped into a final shuttle tanker. All buoyancy tanks were emptied and all valves, watertight hatches and doors were shut to prevent flooding. With these functions completed, Shell began the long, and as it turned out fraught, process of trying to get rid of the Brent Spar.

Chapter 2

Decommissioning and the BPEO process

DECOMMISSIONING NORTH SEA STRUCTURES: THE EXTENT OF THE PROBLEM

Decommissioning redundant offshore structures, i.e. getting rid of them at the end of their working lives, is one of the most pressing problems facing the offshore oil and gas industry worldwide, and particularly in the North Sea. Not unreasonably, having invested millions of pounds in the structures in the first place, the operating companies would like to re-use them or sell them to another company to use elsewhere. But this will rarely be a feasible option. The advance of technology and exposure to the hostile marine environment will usually have rendered the structures out of date, too expensive to maintain, or both – as in the case of the Brent Spar. With limited recycling value, the cheapest way of getting rid of most structures would undoubtedly be to simply topple them where they stand, or drag them into deep water and sink them. It was therefore understandable that many of the media reports on the Spar picked up on Greenpeace's concern that its deep-sea disposal would create a precedent for the disposal of other structures in the North Sea. But many of the numbers mentioned in this context were simply ludicrous. Although the best reports pointed out that there are many different sorts of structure in the North Sea fields, and that deep-sea disposal was not a likely option for most of them, there was nevertheless a great deal of confusion, with many suggestions that 400 or more of these structures might end up in the deep ocean. We shall look at whether the Brent Spar's disposal might indeed have created a precedent in a later chapter. But first, let us look briefly at the problem in general, its scale in the North Sea, what the 'rules' were in 1995, and what Shell had to do specifically to obtain a licence to dispose of the Spar. The following figures are from a variety of sources, but mainly from UKOOA and from the Third Report of the House of Lords Select Committee on Science and Technology, published in February 1996.

First, as far as marine disposal is concerned, whether in deep or shallow waters, the problem was a finite one since it would definitely not apply to structures installed in the future. It was agreed between all the companies and the licensing authorities that no installations would be emplaced in the North Sea after January 1998 unless they had a 'cradle-to-grave' plan for their safe disposal on shore. There was a minor caveat to this statement which we will consider a little later, but essentially marine disposal simply could not be considered for any future installation. So what about the existing structures?

Since 1967 when North Sea oil and gas exploration began, about 440 steel and concrete installations have been emplaced, about 200 of them, together with almost 7,000 kilometres of pipeline, in the United Kingdom sector. These structures have a combined total mass of almost eight million tonnes. They include steel and concrete fixed platforms, floating production platforms, sub-sea production systems, all of them what most of us would think of as 'oil drilling rigs' of one sort or another, and finally there are the offshore storage and/or loading units of which the Brent Spar was one. Each is unique, though they can be grouped into a relatively small number of broad categories.

The platforms in the relatively shallow waters (to about 30m depth) of the southern North Sea, about 150 of the UK sector installations, are mostly so-called 'steel-jacketed platforms', weighing between 1,000 and 5,000 tonnes. A similar number stand in other national sectors of the North Sea, and all consist of a 'topsides' section weighing up to about 2,000 tonnes and housing the main working areas and accommodation, standing on a supporting framework typically fixed to the seabed with steel piles and concrete grouting. There is no question of these structures being disposed of in the sea. Because of their relatively small size, they present no great engineering problems and it was already agreed that they would all be completely removed following International Maritime Organisation (IMO) Guidelines.

The fifty or so installations in the deeper (up to 200m deep) and climatically more extreme northern areas of the UK sector of the North Sea, and about the same number in the Norwegian sector, all tend to be bigger, heavier and stronger than those in the southern parts. For example, the North Cormorant Platform, a steel-jacketed structure basically similar to many in the southern section, weighs

20,000 tonnes. 'Gravity based' structures, like three of the four production platforms in the Brent field, may also be grouted in position, but they remain in place mainly because of their immense weight and the suction created with the seabed during installation of their concrete bases. They are generally even heavier than the steel structures. The Beryl Alpha Concrete Gravity structure, for example, has steel topsides weighing 30,000 tonnes, while the total installation weighs in at 522,000 tonnes. It stands in 119m of water and can store almost a million barrels of oil. But there are even bigger structures; the Shell concrete gravity base platform for the Norwegian Troll gas field is truly immense, weighing more than one million tonnes and standing in 300m of water! It is amongst these very large structures that the problems of decommissioning and disposal are most acute. So what was likely to happen to them?

First, in the vast majority of cases the topsides of these installations, roughly the bits sticking out of the water, would be cut off and transported ashore for re-use or recycling. There are good pragmatic reasons for this. For one thing, the operation is relatively simple, and in any case many of the structures are far too useful or valuable to be thrown away. Second, where they stand on massive concrete bases it is widely recognised, even amongst environmental groups, that removing them would probably be extremely difficult, if not impossible. So they were likely to be left where they are.[1] While this may be regrettable, there seems to be no other practical solution and, in any case, concrete is relatively inert so the environmental impact is not likely to be great. Finally, the bits in between the bases and the topsides were likely either to be taken ashore or toppled, that is cut, possibly explosively, at some point beneath the surface and allowed to fall onto the seabed alongside the bases. There was already national and international legislation in place governing the clearance required above the remaining stumps, and accurate logging of their positions, to ensure that they did not become navigational hazards. So although there were understandable objections to this scenario from environmental and fishermen's groups, none of these solutions involved deep-sea dumping. In fact, when you remove these various possibilities from the total inventory of North Sea installations, you are left with a very small number, perhaps a dozen or so, of special installations for which deep-sea disposal was ever a real possibility. The Brent Spar was the first, and possibly the last! So what did Shell have to do to obtain permission to dump the Spar in the sea?

To dispose of any significant waste in the UK, or in areas under

UK jurisdiction, the operator must obtain a government licence. In this case, the disposal was to take place in the UK's Exclusive Economic Zone or EEZ, roughly an area of sea extending 200 nautical miles off the UK coastline. EEZs are subject to a raft of international laws and agreements which are dealt with in Appendix 1. But these are the concern of governments, and from Shell's point of view the requirement was to obtain the necessary licence from the DTI who, in the case of disposals in Scottish waters as this one was to be, delegate the licensing function to the Scottish Office Agriculture, Environment and Fisheries Department (SOAEFD).

In order to get the licence Shell had to comply with the BPEO principle, established originally in 1976 by the Royal Commission on Environmental Pollution to govern all aspects of the discharge into the environment of potentially harmful substances. This represented an important change in the official UK attitude towards waste disposal. Previous legislation had been based on the BPO or Best Practicable Means principle which had its origins in the nineteenth century. But the BPO system contained no direct reference to environmental impact. Now, for the first time, the environment was to be a central issue in deciding whether or not to allow a potentially polluting activity.

But exactly how the BPEO system was to operate was not spelt out in the original Royal Commission report. Instead, the practicalities of the system developed gradually and were eventually the subject of the Commission's Twelfth Report published in 1988. Here, the BPEO system was described as 'a procedure that would lead, if properly implemented, to reductions in environmental pollution and to improvements in the quality of the environment as a whole'. Because the procedure should be open, the interpretation of 'practicable' would be subject to scrutiny and the risk of lax environmental practices would be reduced. The Commission listed a series of factors which should be taken into account during the BPEO procedure, ranging from the need to seek alternative options diligently, to identify the origins of all data used so as to be able to assess its reliability, to present scientific evidence objectively, and to consult widely with interested parties. To address these factors the Commission identified a series of specific steps through which any licence applicant would have to go. Let us see how Shell did in the case of the Brent Spar.[2]

After the Greenpeace occupation, and in the harsh light of the resulting publicity, Shell made considerable play of the fact that since the Spar had been decommissioned some thirty separate reports on it had been prepared for Shell, many by specialist consultants.

Although few, if any, of these reports were confidential, many of them are relatively inaccessible to most people because they are quite detailed and technical and difficult for the non-specialist to understand fully. But their results and conclusions are broadly summarised in two fairly short key documents which were submitted to the Government at the end of 1994 as the final submission for the disposal licence. Both produced for Shell by Rudall Blanchard Associates Limited (1994a; 1994b), these are the 'Brent Spar abandonment BPEO assessment', setting out the arguments leading to the conclusion that the deep-sea disposal was the best one, and the 'Brent Spar abandonment impact hypothesis', which attempts to identify the nature and extent of the environmental impacts that might be expected. Using these documents we can work through the Royal Commission's recommended steps and assess how well or badly Shell did at each stage.

Step 1: Define the objective in terms which do not prejudge the means by which it is to be achieved

This was fairly obvious – to get rid of the Brent Spar. The BPEO document is very clear on this, pointing out that refurbishing it in 1990 to continue its working life would have cost more than £90 million and that the Spar would have been non-operational for 2–3 years. Having been taken out of commission in October 1991 and effectively 'mothballed', by the time the BPEO document was written the Spar was officially described as a 'not normally manned installation' and its Certificate of Fitness did not allow it to be used for oil storage even if Shell had wanted to use it in this way. In order to renew the certificate a full structural integrity survey and refurbishment programme would have been required. So not only was the Spar useless to Shell, but it was costing them money to leave it where it was, and would soon have cost them a good deal more. No wonder they wanted to be rid of it.

Steps 2 and 3: Identify all practicable options and analyse them, particularly to expose advantages and disadvantages for the environment

A study for Shell by McDermott Engineering (Europe) Ltd in 1993 identified thirteen different options for abandoning or re-using the Spar. This list was screened by Shell and reduced to six considered to warrant further study, two for its re-use and four for its abandonment one way or another.

In fact, only one of the shortlisted options was truly for the Spar's re-use, its refurbishment in the hope of finding an alternative buyer. Not very surprisingly, no buyer was found and this possibility was dismissed at an early stage.

The second so-called re-use option was simply to carry out the minimum amount of maintenance at its original worksite to stop the Spar from deteriorating. Since Shell had no intention of using the structure again, this would have been only a delaying action and not a final solution. Interestingly, the BPEO document cited, as a reason for at least considering this option, 'to delay abandonment and combine it with other field abandonment programmes'. No-one seems to have picked up this point which might have been interpreted as indicating that Shell thought that some abandonment package deal was on the cards for the Brent Spar and other decommissioned structures. So much for the case-by-case argument!

The remaining options on the McDermott list were in-field disposal, vertical dismantling, horizontal dismantling and, finally, deep-sea disposal.

The in-field disposal idea was to sink the Spar at, or very close, to its original worksite. This was by far the simplest and cheapest option, and also the safest since it would involve minimal exposure of the workforce to hazardous operations. However, it would have meant leaving a major obstruction on the seabed and the release of the Spar's contents into the North Sea. Shell clearly realised that this would have been totally unacceptable to the authorities, the environmentalists and the public, and the plan was not considered further.

Two of the remaining options were for the Spar's ultimate breaking up and disposal ashore, differing only in the way it was to be transported and dismantled. Vertical dismantling would have involved towing it intact to a sheltered deep-water site to be partially decontaminated and cut into horizontal slices which would be transported ashore for final mechanical breaking and disposal. Since this would not have involved reorientating the Spar from the vertical to the horizontal, always emphasised by Shell to be a particularly difficult and hazardous operation, McDermott considered this option to be less complex than horizontal dismantling. But the only UK site identified by McDermott as suitable for the slice-cutting operation was Loch Kishorn on the northwest coast of Scotland, and even this locality has a draught restriction of 80m compared with the Spar's normal 109m. The required deballasting would consequently have made the operation considerably more complex and increased the risk of something going wrong. According to the BPEO document, other possible deep-water European sites were eliminated as possible destinations because of 'regulatory constraints', that is, Shell did not think it would be allowed to take the Spar to such non-UK localities. In view of the readiness with which Norway subsequently allowed them to park the Spar in Erfjord for more than two years, this conclusion seems to have been somewhat premature.

But Shell and its advisers thought there were other disadvantages in the vertical dismantling option. First, there would be considerably greater exposure of the workforce to hazardous operations than in the case of horizontal dismantling. Second, there was a high perceived risk in this option of things going badly wrong, at worst of total loss of control over the Spar and its consequent sinking and break-up, possibly in environmentally sensitive inshore waters. Finally, the estimated cost of some £44 million was considered too high. All in all, and particularly because of the failure to find suitable deep-water sites for its execution, vertical dismantling was considered to have no advantages over horizontal dismantling and was accordingly dismissed. Again, in view of the final solution, this decision seems a little questionable now.

Horizontal dismantling would have involved removal of the topsides section, repair of the damaged storage tanks, rotation of the buoy to the horizontal, its transfer to a cargo barge and its transport ashore for final break-up. A detailed assessment of these operations concluded that this would be by far the most technically difficult option, with reverse upending being the most critical stage. After the removal of the 1,570-tonne topsides section, the repair of the tanks and the installation of the reverse upending control system of valves, pumps and pipework, the Spar would have to be towed to a suitable site some 80km north of its working position. Here, the buoy would be placed in a pre-installed mooring system accompanied by a heavy lifting vessel, a work barge and a tanker. The contaminated water in the storage tanks would now be pumped into the tanker and replaced by an inert gas mixture of nitrogen and carbon dioxide, thus deballasting the buoy. By varying the rates at which the individual tanks were emptied of water the Spar would be gradually tipped to the horizontal position. But because of the thinness of the tank walls this would have to be done extremely carefully to avoid exposing them to excess differential pressures which might cause them to collapse. And the still-vulnerable repaired tanks would have to be orientated out of the water as much as possible. Reverse upending would clearly be a very ticklish operation in sheltered inshore waters. In the open sea it would be even more difficult and, since it was expected to take between five and ten days to complete, would be extremely vulnerable to vagaries in the weather.

Once the Spar was horizontal the rest of the operation would be comparatively simple, though to a non-engineer still presenting horrific problems. With the pumping equipment removed it would be impossible to remove any water that leaked into the tanks during the subsequent tow to a sheltered site for transfer to a submerged cargo vessel. Scapa Flow was identified as a possible transfer site, where two 2,000-tonne capacity shearlegs would be needed to lift the Spar sufficiently to manoeuvre the semi-submersible beneath it. This was deemed necessary to reduce the 15m draught of the horizontal Spar so that it could be towed to a suitable dockside for dismantling and transport ashore. But this was also partly because only UK final destinations were considered. Again, in view of the wide range of possible break-up locations subsequently put forward, it seems reasonable to suggest that the BPEO process did not consider all the possible options.

Once loaded onto the semi-submersible, the Spar would be transported to the final break-up and disposal site where the tanks would be cleaned and the ballast removed before the carcass was cut into relatively few large lumps. Even now it would have to be treated fairly carefully. Those parts which had been in contact with crude oil would be assumed to have become coated with the infamous LSA radioactive scale, which would have to be removed and disposed of in accordance with the local regulations. Now, at last, the remains of the hull could be cut into manageable sections and offered for sale as scrap or otherwise disposed of.

Despite the obvious engineering complexity of this option, the individual operations involved were considered not to be basically different from those relatively routinely undertaken by the offshore industry. The horizontal dismantling option was therefore kept 'on the cards' for comparison with the final option, deep-sea disposal.

Deep-sea disposal was assessed as the technically simplest next to the rejected possibility of sinking the Spar close to where it had worked. It would have involved only cleaning up the topsides, emplacing explosive charges and a relatively long tow to the disposal site, at that time unspecified but assumed to be one of the three North Atlantic sites identified by SOAEFD. This option could be carried out with or without the topsides in place. But since their removal, it was claimed, would simply increase the risks to the workforce without significantly reducing the environmental impact, this option was not considered further. Instead, the topsides would simply be cleaned up before explosive experts emplaced the charges that would be used to sink the Spar, using enough to ensure that all the buoyancy tanks would be opened to the sea even if some of them failed to detonate. Once the anchor chains had been disconnected, two of them would be used by tugs to tow the Spar to the disposal site, while the others would act as emergency towlines.

When the BPEO documents were published, three possible disposal localities were still under consideration: one in the Maury Channel about 500 nautical miles west of the Hebrides, one in the Rockall Trough, and finally the one ultimately selected, the so-called North Feni Ridge site about 150 nautical miles northwest of the Scottish mainland (figure 4). Each was within the UK's exclusive economic zone and, the BPEO document stated, had been the subject of a 'detailed survey...undertaken by Shell Expro and SOAEFD to confirm the "suitability"...and to provide baseline environmental data for future monitoring programmes'. But as we shall see, the results of these questionable surveys were not yet available; since no final choice had been made, the BPEO study had to consider all of them. Depending on which site was chosen, the tow was expected to take between fifteen and twenty-five days. Finally, once the Spar reached the disposal site the tow lines would be released mechanically, the tugs would stand clear and the explosives would be detonated by remote control.

Steps 4 and 5: Summarise the results of the evaluation concisely and objectively, and select the preferred option on the basis of the environmental impacts and associated risks, and the costs involved

Since all the other possibilities had been rejected at an early stage, only horizontal dismantling and deep-sea disposal were evaluated in any detail against the criteria of engineering complexity, risk to the workforce, cost, acceptability to the 'authorities and other interested parties', and, finally, environmental impact.

On engineering complexity there was no contest. Deep-sea disposal would be relatively simple, the most difficult part being the tow to the disposal site, with some possibility of 'unplanned events' mainly resulting from inclement weather. But this could be minimised by undertaking the tow in the summer months. In the event, of course, the Spar was towed without mishap not only the 400 or so nautical miles to the abortive disposal site, but in addition almost twice as far back to Erfjord.

In contrast, horizontal dismantling would be an extremely testing operation, with several of the eighteen distinct stages identified being quite difficult engineering tasks and carrying significant risks of 'unplanned events'. However, some of the technical problems, such as those associated with supporting the Spar on a semi-submersible, were undoubtedly exacerbated by the limited range of final disposal destinations that had been considered.

The health and safety risks to the workforce were also considered to be much greater for horizontal dismantling, involving as this would a much larger number of workers in many potentially hazardous activities. The risks of a fatality were assessed as six times higher than with the deep-sea disposal option.

Despite the relative simplicity of the deep-sea option, even throwing the Spar away would, according to Shell's figures, have cost it a massive £11.8 million. But the estimate for horizontal dismantling was a staggering £46 million. Although this figure was challenged by Greenpeace and other opponents of Shell, the Royal Commission's recommendations state explicitly that financial considerations 'should not be overriding' and, in any case, few outside observers on either side seemed to be particularly influenced by the costs one way or the other.

So far, while some would argue that Shell and its advisers had shown somewhat limited imagination in identifying disposal options, particularly the range of possible dismantling sites and even the possibilities for the Spar's re-use or recycling, few would accuse them of not having carried out a reasonably adequate comparison of the engineering, workforce safety risk and cost criteria. Now let us see how they fared with the remaining two BPEO criteria, consultations with interested parties and environmental impact.

According to the BPEO document, discussions on the proposed abandonment plan for the Spar were initiated between Shell Expro and the Department of Trade and Industry in 1992, with the implication that the DTI was aware of, and presumably had an input to, the way the process was conducted from that point on. This is important since, as we have seen, the DTI was the main licensing authority. In Chapter 4 we will look at some of the correspondence between Shell and the DTI when we consider whether the deep-sea disposal of the Spar would have created a precedent.

Under the Petroleum Act of 1987, the UK legislation governing the decommissioning of all offshore oil and gas structures, Shell were also required to consult a number of organisations ranging from the Scottish National Heritage, the Joint Nature Conservancy Council, a variety of professional fishermen's associations and British Telecom International. These bodies were apparently appraised of Shell's preferred option of deep-sea disposal during the first quarter of 1994. Finally, 'detailed discussions' were held with SOAEFD, Her Majesty's Industrial Pollution Inspectorate and the Health and Safety Executive, while the Ministry of Defence, the Hydrographer to the Navy, the Crown Estates Commissioners for Scotland and the Department of Transport were also informed of the abandonment plan. In other words, Shell complied with all the regulations as far as consultation was concerned, and since no objections were raised it obviously thought it was home and dry. But despite the Royal Commission's recommendation that there should be 'appropriate and timely consultation with people and organisations directly affected', there is nothing in the BPEO process requiring that the preferred option should be acceptable to a wider public, or even that the intentions should be widely publicised. In retrospect, it is easy to see that Shell would have been wise to publicise its disposal options – all of them, not just the deep-sea one – much more widely and at a very early stage. Had it done so, Shell might not only have discovered the strength of feeling against the deep-sea option, but also have complied more adequately with the Royal Commission's recommended sixth step, having the preferred option scrutinised preferably by some totally independent body. As it was, there was no mechanism in place to ensure that this was done – and Shell did not do it.

Quite apart from getting the technicalities checked, and particularly those concerning environmental impact, as we will see later, Shell clearly should have targeted the environmental pressure groups, including Greenpeace, in an attempt to attract its support rather than its implacable opposition. Why didn't it do so? We don't know, but we can think of only three possible explanations. The first is simply because it didn't have to. Not only, as we have seen, does the BPEO process not require this, but nor did the DTI guidance notes on decommissioning offshore structures. Apparently, when the guidance notes are revised they may include a recommendation for such wider dissemination of information (see Coleman 1997), but nothing of the sort was in place at the time of the Brent Spar. As we will see in Chapter 7, the 'way forward' and 'dialogue process' put in place by Shell after its capitulation is a self-imposed, laudable, but belated (and some would say rather excessive) attempt to make up for this earlier deficiency. If this was the reason, then both Shell and the DTI showed a remarkable lack of foresight. But the two other possible explanations are much less charitable. Was Shell so confident of the weight of argument in their BPEO study that it thought that the environmentalists would not or could not raise any valid objections to it? Less credibly, did Shell think that the deep-sea disposal plan would sail through the licensing process and its execution so smoothly that no-one would notice? But whatever the explanation for the lack of public information, if not consultation, how wrong it was! Apart from totally underestimating the vigilance, tenacity and bottle of Greenpeace, Shell apparently failed to realise that the environmental impact assessment within the Brent Spar BPEO process would come under particular scrutiny.

From the point of view of public acceptability, the E in the BPEO process is clearly crucial. Indeed, as part of the process an applicant for a licence is required to carry out a specific study of the potential environmental impacts of the preferred option under a range of scenarios. In the case of the Brent Spar the results of several reports on this topic were brought together in the Impact Hypothesis document. It contained a number of errors which confused an already murky situation. To understand where these errors came from we have to go back to a much earlier stage in the decommissioning procedure and look at the way 'science' was used on both sides of the debate.

Chapter 3

The scientific debate

Shell abandoned their deep-sea disposal plan for the Brent Spar for a complex mix of socio-economic reasons, but certainly not because they had 'lost' the scientific argument. Similarly, Greenpeace succeeded in mobilising opposition to the plan by appealing to a strong pro-environment and anti- big business sentiment in the general public, and not because they had 'won' on science. Nevertheless, both sides used science as an important part of their respective cases: Shell to try and demonstrate the environmental acceptability of the deep-sea disposal option, which happened also to be the cheapest and technically easiest; Greenpeace to negate Shell's arguments even though they subsequently claimed that it was the principle and precedent elements of the marine disposal to which they objected, rather than the specific environmental impact of the Spar itself. Although science *per se* was not therefore an important factor in the eventual outcome, the apparent disagreements and confusion in the scientific arguments certainly supported the widespread feeling that the issues were not as clear-cut as Shell and Government spokespersons seemed to imply. In fact, there was so much scientific confusion that in late 1995, Tim Eggar, the then Energy Minister at the DTI, asked the Natural Environment Research Council to set up the Scientific Group on Decommissioning Offshore Structures specifically to look into these issues. The accuracy of the resulting report (referred to here as the Shepherd Report after its Chairman, Professor John Shepherd of the Southampton Oceanography Centre) has not been questioned by any of the interested parties and will therefore be used as a standard against which to compare the statements and claims made by the other groups. To examine the scientific arguments we will start with the Greenpeace anti-dumping case, since this was the one that first impinged on the public; we then move to the conclusions of the

Shepherd Group, and finally compare these with those of Shell and its consultants. This will lead inexorably to the much publicised supposed argument between the 'experts', and finally to a sad and embarrassing story about how the site-specific surveys were bungled. The conclusion is that the 'scientific debate' about the Spar was not a rational argument about what is or is not known, but instead was a confused and fruitless series of exchanges mostly based on misinformation and misunderstandings.

THE GREENPEACE CASE

Greenpeace are fundamentally opposed to dumping anything in the oceans, openly admitting that this is based on moral and aesthetic grounds as well as on science. Nevertheless, much of their public opposition to the deep-sea disposal of the Spar was based on science, and particularly on its likely impact on the environment. The case was based initially on a rather good study carried out by an engineer with considerable oil and gas industry experience (Corcoran 1995) and was subsequently supplemented and summarised in a letter from Dr Helen Wallace of the Greenpeace Science Unit and sent to many scientists on 5 July 1995 following a critical opinion piece in the journal *Nature*. This letter sets out the Greenpeace case under the following headings.

There had been no formal inventory of the Spar's contents, so the environmental impacts could not possibly be properly assessed

Greenpeace pointed out, quite correctly, that Shell's data on the contents were largely estimated rather than measured. In retrospect, after the DNV survey had been completed, we knew that the Shell figures were in essence correct, but at the time this criticism seemed very reasonable.

There is a lack of understanding of the deep-sea environment, and it is currently impossible to predict the effects of the proposed dumping on deep-sea ecosystems

This was based on the assertion that the dispersal of pollutants (presumably such as those in the Spar) in ocean sediments is difficult to predict and, more crucially, that the likely effects on deep-sea organisms are unknown. This is true. It was, and is, impossible to predict the exact consequences of the deep-sea dumping of the Spar. Knowledge of deep-ocean science is still very fragmentary and oceanographers are continually surprised by deep-sea phenomena that they never dreamed of. It is also true that nothing is known of the sensitivity of deep-sea animals to toxic chemicals and it is currently difficult to see how such information could be obtained. On the other hand, although deep-sea animals, being accustomed as they are to a relatively constant and 'clean' environment, might be expected to be rather more sensitive than their cousins from the rough-and-tumble of shallow waters, it is inconceivable that their physiological processes are so different that they are orders of magnitude more susceptible. But you do not need to be able to predict the *exact* consequences, nor to know precisely how sensitive the animals are, to produce a confident ballpark estimate of the effects of dumping the Spar based on limited available knowledge. As we shall see, the Shepherd Group did just this – and came to the conclusion that the effects would be both extremely local and relatively insignificant.

The documents supporting Shell's licence application are highly conjectural. They contain numerous unsubstantiated assumptions, minimal data and extrapolations from unnamed studies

Greenpeace was able to point to a number of questionable statements, including assumptions about the contents and their possible toxicological effects referring back to the first two objections. But Shell also included assumptions about the fate of the Spar as it sank, concluding that it would not break up in the water column, nor disintegrate catastrophically when it reached the bottom. Instead, it was suggested, it would broadly retain its integrity on the seabed so that the contaminants would be released slowly over about one thousand years and the resulting particulates would settle evenly within a very short distance. Greenpeace questioned these assumptions, as did the engineers on the Shepherd Group, who concluded that the Spar might well break up in mid-water and almost certainly on impact with the bottom. But despite the resulting rapid release of the contaminants, the group concluded that their serious effects would be extremely local.

As we shall see, Greenpeace made considerable mileage out of more serious shortcomings in the licence documentation when these came to light some weeks after the Wallace letter, but they were not part of the original case.

Now let us move to the conclusions of the Shepherd Group.

THE SHEPHERD GROUP FINDINGS

The Group's First Report, published in May 1996, is a detailed 76-page document which comments on the procedure followed in the licensing process, on the engineering aspects and even on the monitoring programme that would have been necessary if the Spar had been dumped. But the crucial part is that concerned with the likely environmental impact, on which it contains the following specific conclusions.

First, that the physical effects would be restricted to 'a relatively small area (less than 1 square km)', and that the effects of toxic materials in the Spar similarly 'would be confined to the immediate vicinity of the wreckage'. That oil released into the water column would 'result in lethal effects on pelagic organisms within a volume of 0.2 cubic kilometres and sublethal effects within a larger volume of 2 cubic kilometres'. That the oily water released when the Spar reached the bottom would affect volumes ten times less than in mid-water. Oily sludge released on the bottom would also kill the seafloor animals, but over only about half the area affected physically. Recovery from these effects would begin within months, and would be complete within 2–10 years. In summary, said the report, 'The global impacts on the environment and on human health of the deep-sea disposal of a structure such as the Spar would be very small, roughly equivalent to the impacts associated with the wreckage of a fairly large ship'.

Nowhere did the Report endorse the deep-sea disposal option as the BPEO, stating categorically that 'Nothing in this report should be taken as promoting the deep sea disposal of decommissioned off-shore structures, or of any other wastes'. But neither did it dismiss the deep-sea option or endorse any others – and for two good reasons. First because, being made up entirely of marine scientists and engineers, the Group did not feel it was qualified to comment on the non-marine options, and second, because it recognised that the scientific issues are not the only ones to be considered in coming to such a conclusion. So all sides in the affair were able to welcome the report and use it selectively to support their own point of view.

However, as far as the scientific debate about the likely environmental impacts is concerned the report was unequivocal; the impact would have been relatively small and local, and certainly not of the scale implied by some of the Greenpeace statements. Several scientists, independent of both Shell and Greenpeace, said as much in the summer of 1995. But the authoritative Shepherd Report did not appear until almost a year after the original arguments had raged in the media during, and particularly shortly after, the Greenpeace occupation. At the time this argument was inconclusive, simply confusing an already confused public. To understand where at least some of this confusion arose we must now look at Shell's scientific case, and specifically at what its consultants thought would be the likely scenario if the Spar were to be sunk in the deep ocean.

THE SHELL CASE

The Shell case was encapsulated in the two relatively brief Rudall Blanchard documents which were submitted to the Scottish Office to obtain the disposal licence. They were eventually published in December 1994, that is, after the site-specific survey cruise (see below) but before the licence was granted. Indeed, it was principally consideration of these documents that prompted the UK Government announcement, in February 1995, that they expected to issue a licence. Like the later Shepherd Report, the Rudall Blanchard documents came to the conclusion that the environmental impact of the Spar dumped in the deep ocean would be relatively small and rather local, though unlike the Shepherd Report, of course, they went on from this to conclude that deep-sea disposal was the *best* option. But they also contained a number of questionable statements that were at the root of the later confusion. These included the assumption that the Spar would broadly maintain its integrity on its final journey to the seabed so that the enclosed contaminants, and its structural materials, would be released relatively slowly over many years. Several commentators questioned this conclusion, including the engineers on the Shepherd Group, suggesting that the Spar might well break up during its descent and would almost certainly do so when it hit the bottom. While this would have a significant effect on the rate of release of the contaminants, it would not seriously affect the overall conclusion and did not raise much public comment. In contrast, some erroneous statements in the Rudall Blanchard documents about the characteristics of the deep-sea environment were exposed to the full critical glare of public scrutiny. So where did they come from and what did they say?

The problem was that the Rudall Blanchard documents were based largely on two environmental impact assessments carried out for Shell Expro by environmental consultants long before the potential disposal sites had been identified (see Metocean 1993; AURIS 1994). In the absence of any guidance on the likely disposal sites, both of these assessments had concentrated their appraisal on abyssal plain sites, that is in depths in excess of about 4,000 metres.[1] At such depths, as both reports said, though often rather unclearly, the near-bottom water currents are generally very sluggish, the seabed animal communities are 100 to 1000 times less abundant than on the continental shelves, and fish, in particular, are far too sparse for any possibility that they will be exploited commercially in the foreseeable future. Furthermore, while these communities are far more species-rich than was thought even ten years ago, they are much less diverse than those on the continental slopes. But the UK has absolutely no abyssal localities within its EEZ. And any site for which the UK Government could contemplate giving a disposal licence must be within its EEZ. So when, in mid-1994, the Environmental Protection Section of SOAEFD's Marine Laboratory in Aberdeen, one of the licensing authorities, selected the three sites to be considered, they were all much shallower than any abyssal site, the ultimate 'winner' being more than 2 kilometres shallower! But neither Rudall Blanchard nor Shell seemed to notice this. And if anyone else did, including SOAEFD, they did not apparently try to correct the error.

Consequently, three specific statements in the BPEO and Impact Hypothesis documents were wrong, particularly as applied to the North Feni Ridge site finally selected. First, that the near-bottom current velocities are typically 10cm/second or less, whereas published data showed that they could often reach 50cm/second or more. (Incidentally, these relatively high current speeds at the chosen site also result in a better sorted, that is coarser and harder, bottom sediment than is normal in abyssal localities and therefore an even greater chance that the Spar would have disintegrated when it hit the seabed.) Second, that no commercial fishing activity was likely in the near vicinity, whereas fisheries already existed to depths of almost 2,000 metres around the nearby Rosemary Bank. Third, that only a 'small range of species' would be found at the disposal site, whereas abundant recent data suggest that the animal communities on the deep seafloor as a whole are species-rich, and that this is particularly so at mid-slope depths, precisely those chosen for the disposal of the Spar!

If Shell and its advisers had wanted to shoot themselves in the foot and provide Greenpeace with ammunition with which to question their scientific arguments, they could hardly have done better.

THE SCIENTIFIC ROW

As we have seen, although the Shepherd Group subsequently concluded that the chosen North Feni Ridge site was probably not the most suitable that could have been chosen within the criteria used, even including the restriction to the UK's EEZ, it did not believe that the disposal of the Spar at the North Feni Ridge site would have caused significant environmental damage. So despite the errors in the BPEO and Impact Hypothesis documents, the similar conclusion reached in these documents was also probably correct. Greenpeace would still have opposed the deep sea disposal even if the BPEO documents had contained no mistakes, of course, but the errors naturally allowed doubts to be cast on the quality of the scientific case and Greenpeace took full advantage of this.

Having finally seen the two licensing documents in July 1995, two acknowledged experts on the Rockall Trough and the proposed disposal site, Drs John Gage and John Gordon from the Scottish Association for Marine Sciences Laboratory at Oban, wrote to Shell, Greenpeace and the press pointing out the mistakes (see *New Scientist*, 26 August 1995; *Scotland on Sunday*, 10 September 1995).[2] Greenpeace welcomed these criticisms with open arms and paraded the two Johns at every opportunity, including a fringe meeting of the British Association's (BA) annual conference in Newcastle in mid-September. By this time they had been joined by other dissenting scientific voices including Dr John Lambshead from the Natural History Museum, who, at a lecture at the BA meeting earlier in the week, had suggested that far from housing a 'small range of species' the proposed dump site might be richer in terms of species, particularly nematodes, than a tropical rainforest. This was widely reported and made particularly good copy since, only a few weeks previously, Tim Eggar the DTI Minister had been reported as saying that the deep-sea disposal of the Spar would cause 'the death of a few worms'. This was a misquote of something that one of us (ALR) had been reported as saying in support of the claim that the Spar would cause relatively little damage, though emphatically not based on the Rudall Blanchard documents.[3] So not only was there reason to doubt the adequacy of the scientific case in the BPEO document, but there also seemed to be disagreement between the country's so-called deep-sea biology experts. The resulting publicity must have been particularly confusing to any onlooker trying to understand the scientific arguments, since the scientists directly involved were all employed directly or indirectly by the Natural Environment Research Council (NERC) and might reasonably be expected to speak with more or less one voice.

In fact, it even confused NERC's Chief Executive, Professor John Krebs, who, in early September 1995, wrote to John Gage and Tony Rice for clarification of the apparent disagreements, particularly in relation to the likelihood of contaminants being brought back to the surface, the biodiversity of the proposed dump site and the nature and degree of connection between the surface layers and the deep ocean. He was reassured that there was no fundamental difference of opinion on the science. The only 'argument' was that while Tony Rice felt that the available knowledge of the deep sea, limited as it is, was sufficient to assess the scale of the Spar's impact with some certainty, John Gordon, and particularly John Gage, were being rather more cautious. Furthermore, Professor Krebs was told, the errors in the BPEO documentation pointed out by Gage and Gordon were irrelevant to this general point.

But the general confusion about the scientific arguments continued to grow throughout September, fuelled by the events at the British Association (BA) meeting. Finally, as we have seen, in October Tim Eggar asked Professor Krebs to set up an independent committee of scientists to look into the matter – ultimately the Shepherd Group. When its report appeared it included some relatively mild but pointed criticisms of the process which had led to the issue of the licence, including the fact that the errors in the documentation might have been corrected if the process had not been so secretive until a relatively late stage. It did not say so explicitly, but the implication was that if the documents had been seen by someone familiar with the issues involved, particularly deep-sea oceanography, the errors would have been spotted. The report also pointed to shortcomings in the science associated with the final part of the BPEO process, selecting and surveying the potential disposal sites with a view to monitoring the effects of the disposal if it had taken place. Let us now look at that.

SITE SELECTION AND MONITORING

Apart from expressing doubts about how the three potential disposal sites had been selected in the first place, the Shepherd Report was critical of the surveys that had been conducted to make the final choice between them, suggesting that the specifications for these surveys 'were not fully appropriate for deep-sea disposal'. Moreover, the report went on to say that 'When the site survey contract was placed, the requirements of the original site survey specification were relaxed'. This is fairly damning stuff which was not picked up when the report was originally published. What was it referring to?

Any pre-operation dump site survey, on land or in the sea, has two main purposes: first, to check on the assumptions that formed the basis of the site's selection in the first place; second, to provide baseline data against which to compare conditions after the disposal, to check whether the forecast effects were correct or not. The data collected must clearly address these purposes. So what happened in the case of the Spar?

Sometime early in 1994 three potential disposal sites were identified by SOAEFD (see figure 4). The criteria for their selection were extremely general and based on the Oslo-Paris Convention, with the crucial overriding criterion being that any site should be within the UK's EEZ. There is no indication that important environmental characteristics for the fate of a dumped Spar, such as the nature of the bottom or the speed of the near-bottom water currents, had been considered at all. The Shepherd Report pointed out these shortcomings, but was even more concerned about the subsequent site surveys and the interpretation of the results. Although couched in fairly diplomatic language, it was quite critical of the role of SOAEFD, who had been the technical advisers and had specified the work to be undertaken in the surveys. In particular, the report implied that insufficient attention had been given to the biology at the sites and the nature of the bottom sediments.

The invitations to tender for the survey work, accompanied by a 'scope of work' document describing what was to be done, were sent out by Shell Expro in May 1994, a year before the intended disposal operation. The surveys were to be conducted between July and October 1994, but the long-term current meters which would have to be deployed would not be recovered until the following spring. This inevitably meant that the current data could not be available until after the BPEO documents had been submitted, and very shortly before the intended disposal itself. In the event, the situation was much worse. Although the survey results would undoubtedly have been available to the decision makers in time for the licence to be issued in early May 1995, they were not generally available for wider scrutiny until February 1996, while the sections dealing with the biology of the larger organisms are not available even now! But ignoring this curious timing, let us return to the survey itself.

The contract to undertake it was eventually awarded to Britsurvey, a division of Svitzer Ltd of Great Yarmouth, who in turn subcontracted specific parts of the work to a number of other specialists. The survey was conducted during August and September 1994, with the current meter moorings being recovered in March 1995. The final report to Shell (Svitzer Ltd 1996) appeared in February 1996, after the Spar had already been moored in Erfjord for several months. It contains vast amounts of information of doubtful relevance, and little that would have been really valuable in comparison with subsequent surveys. This probably doesn't much matter since it will never be needed for its original purpose. Unfortunately, in retrospect there is a much more worrying side.

Among the recipients of the tender invitation was the NERC Institute of Oceanographic Sciences Deacon Laboratory (IOSDL, now part of the Southampton Oceanography Centre). The Institute responded on 16 June 1994, but pointed out that 'because sections of the tender specification are fundamentally flawed we are unable to offer a full tender submission', because 'any environmental survey we participate in must use up-to-date knowledge, methods and equipment'. The letter went on to point out where, in the opinion of the Institute's experts, the survey specification was faulty and offered its help to improve it. There was no reply. When, some six weeks later, the Institute contacted Shell Expro to find out what the situation was it was told that it would not be receiving a contract because its approach had been 'too scientific'!

The significance of this series of events is not that IOSDL, the UK's foremost deep-sea research organisation, did not get the contract, or even that its advice on the survey was ignored. The real significance is that a year later, when NERC scientists, including IOSDL staff, were offering their professional opinions about the Spar affair, various commentators expressed surprise that the NERC had not been involved at a much earlier stage. On several occasions, the NERC's Chief Executive, Professor Krebs, was able to give a categorical 'no' answer to the question as to whether the NERC had been asked for or had offered its advice at any stage in the affair. As we have seen, this was literally true. And what lay behind it was the fact that both in the initial studies of the scientific issues surrounding the Spar, and in the subsequent site-specific surveys, the country's best available expertise was simply not used. No wonder the documentation contained errors and resulted in sterile and inconclusive argument. There has got to be a better way.

Chapter 4

The use and abuse of precedent

From the first Greenpeace press releases after the Brent Spar occupation, and throughout the ensuing debate, the environmental group and other objectors to the deep-sea disposal of the Spar referred time and again to the 'precedent' that would be created if Shell's original plan was accepted. The implication, tacit or explicit, was that if the Spar was dumped it would be taken by the offshore oil and gas industry as *carte blanche* for the similar disposal of other North Sea installations.

In this context, there were frequent references to the 400 or more structures in the North Sea as a whole and the 200 plus in the British sector, again with the implication that all of them were likely candidates for deep-sea dumping. This was, of course, nonsense because most of these structures are relatively small, easy to dismantle to take ashore and the intention has always been to remove them totally from the sea at the end of their useful lives (see Appendix 1).[1] But it was certainly true that for up to a hundred or so of the largest installations, particularly in the northern North Sea, the companies which owned and operated them would wish to consider deep-sea or *in situ* disposal for at least some parts of them among the various options as they were decommissioned. The question was, would the disposal of the Brent Spar in the deep Atlantic have made the use of this option more or less inevitable for these other structures? Greenpeace said 'yes'; Shell, and particularly the UK Government, emphatically said 'no'. Having looked at the evidence fairly dispassionately, we conclude that the answer is 'very probably'.

The Government argument was based on the claim that there was no question of precedent since each application for a disposal licence is dealt with on a 'case by case' basis, that is, without reference to any other application. But this surely could not be literally true. Apart from objecting to marine waste disposal in principle, one of Greenpeace's major arguments against it is that a series of seemingly small impacts, each considered to be relatively insignificant and therefore 'acceptable', might together add up to an unacceptable impact. Any proposal to dispose of waste in the deep ocean, of a Brent Spar or anything else for that matter, must surely consider the current situation, including the effect of any past disposal, particularly if the 'new' disposal is to take place in the same area as an earlier impact.[2] But is this what actually happens?

The Shepherd Group were clearly not too sure. In their general conclusions and recommendations they make the point (NERC 1996: 7, C7) that 'Existing procedures require the case-by-case evaluation of disposal options, which is necessary, but not sufficient, because continued disposals with small individual impact might give rise, by small increments, to an unacceptably large overall impact'. They accordingly went on to recommend 'that assessments of both cumulative and case-by-case impacts be made'.

So how confident are we that the case-by-case approach, espoused by the UK Government and claimed to answer any fears of precedent, works or would work in this rational way, neither tending to adopt past solutions simply because these were apparently acceptable at the time, nor being unduly influenced by what might come after? Frankly, we are not very confident at all. At the time of Shell's capitulation, disposal by sinking at the proposed Brent Spar deep-sea disposal site was being actively considered for at least one other major North Sea installation, though not by Shell. Open discussion of the relevant documentation is precluded by confidentiality clauses. But correspondence between Shell and the DTI that is already in the public domain, or at least in the House of Commons Library, suggests that the UK Government would not have been surprised to receive further applications for deep-sea disposal licences and might well have been inclined to issue them.

This correspondence consists of a series of twenty-four letters between the Department of Trade and Industry and Shell, written between 1992 and 1994 and made available to the then opposition environment spokesperson, Frank Dobson, in February 1996. Mr Dobson subsequently released the letters to Greenpeace who, in late January 1997,[3] maintained that these demonstrated that it was the DTI rather than Shell that pushed initially for the deep-sea disposal option for the Brent Spar (see, for instance *Oil and Gas Journal*, vol. 95, issue 5, 3 February 1997; *Guardian*, 24 January 1997). One of these letters, from Shell to the DTI in August 1992, reveals that the Department had asked Shell to provide 'the relevant factors that might influence a decision to permit dumping of the Brent Spar rather than demolition and scrapping'. Although the exact meaning of this phrase is open to different interpretations, the tenor of the accompanying 'note' stresses the perceived advantages of deep-sea disposal compared with onshore scrapping. Shell certainly seems to have interpreted the DTI request as a requirement for arguments in favour of the deep-sea option even at this early stage.

According to a subsequent letter from Shell in March 1993, the company had already made an 'exploratory proposal to dispose of the Spar by sea dump' as early as October the previous year, but it is not clear whether or not it was encouraged by the Department to make such a proposal. This letter also points out that Shell was hoping to dispose of the Spar 'either in August/September 1993 prior to the onset of the winter season or in May 1994 at the beginning of the summer season'. The following month a 'Brent Spar Abandonment Plan' was submitted to the DTI, who received comments on it from SOAEFD, who had clearly suggested that the Spar's topsides, at least, might be removed for onshore scrapping. Shell's response (14 December 1993) was reasonably bullish, pointing out that it did not see 'any real environmental gain from this option over the sea disposal of the whole after the removal of the environmentally sensitive materials', that 'the feasibility of removing the Spar topsides in-situ offshore has not been established', and that considering this possibility would delay the abandonment until 1995, with consequent increased costs. Furthermore, in response to concerns 'regarding the precedent that would be set by dumping the Spar topsides, Shell pointed out that '15,000 tonnes of topsides being removed [from other] Brent platforms during redevelopment will be scrapped onshore', but that 'in the case of Spar we remain convinced that dumping of the unit as proposed is the correct course of action'.

Up to this point it seems to us that you would be hard pressed to decide from the correspondence which side was pushing the deep-sea disposal of the Spar more strongly, though both clearly thought that it had great merit. But by the end of the year Shell was becoming a little impatient at the slow progress. Its UK Chairman, Dr Fay, had written to DTI Energy Minister Tim Eggar complaining about the delay and had requested the Department's approval for them 'to commence the formal consultation process'. On 25 January 1994 the

DTI wrote to Shell explaining that it was still occupied in ensuring that, in the event of deep-sea disposal of the Spar, 'the United Kingdom's international obligations are fully discharged', and that it was not yet convinced on the question of the prior removal of the topsides or of the complete removal of the concrete anchor blocks. Clearly, the Spar disposal would not take place before 1995 at the earliest. However, two phrases in this letter suggest that the DTI by this time was not only committed in principle to the deep-sea disposal of the Spar,[4] but also believed that the same solution would be applicable to other installations. With regard to the approval for formal consultation, the Department saw 'no difficulty with this as long as it is recognised that such consultations are without prejudice to the final decision on deep sea disposal and that this position is made clear to the consultees'. Does this mean that a year before the final BPEO document and the Impact Hypothesis document had appeared, the DTI had already decided that the Spar would be dumped in the deep sea? It certainly seems so. Moreover, had it also already decided that other decommissioned installations would be disposed of in the same way? 'One of the key issues still to be resolved', the letter said,

> is the presence of LSA and metallic compounds in the Spar and the implications this has for deep sea disposal. This is an issue which is likely to feature in the abandonment of other offshore installations in the future. It is therefore essential that the measures we put in place for disposal of the Spar can be upheld as having addressed all the issues.

It is difficult to read this statement without concluding that the DTI believed that other decommissioned structures would be disposed of in the deep sea, or at least that licences to do so would be applied for. But up to this point there had been no reference to any particular disposal site or sites. Indeed, no specific sites had been selected when the early studies on the potential impact of deep-sea disposal had been carried out, ultimately resulting in some of the scientific confusion. By early 1994, however, SOAEFD had selected the three potential disposal sites within the UK EEZ which were ultimately surveyed later that year. By now, the timetable that would have to be followed in submitting the necessary documentation for a UK disposal licence and to submit the abandonment plan to the Oslo Commission in time for a 1995 summer operation was becoming critical. Consequently, since the results of these surveys would not be available until well into 1995, a letter from Shell to the DTI on 14 June 1994 'assumed that the Environmental Impact Assessment will not be reworked for the different sites, rather a statement of the SOAEFD survey proposals for the 3 possible sites included'. Moreover, Shell also assumed that 'The BPEO will be a separate stand alone document also not site specific', as, indeed, it was.

Nevertheless, the surveys of the potential sites were about to be undertaken at Shell's expense. So in the same letter, Shell wrote that 'It would assist our efforts to pursue industry funding through UKOOA [United Kingdom Offshore Operators Association] if a statement could be provided on the general applicability of the proposed abandonment sites to other oil industry users'. As far as we are aware, Shell did not receive financial support for the site specific surveys from any other sections of the offshore industry. But from the company's point of view, this possibility would have seemed reasonable. If any of the sites were likely to be considered for subsequent disposals, why should Shell bear all of the costs simply because it was the first North Sea operator to consider deep-sea disposal for decommissioned structures? On the other hand, if any other companies were to contribute to these costs, presumably they would have done so only if they were fairly confident that it would have been worth their while to do so; in other words, that they would have been able to dispose of their own decommissioned structures at the same localities. Although this would not have negated the use of a case-by-case approach to subsequent licence applications, it would have suggested that such companies believed that the Brent Spar example could readily be followed. To an outside observer, the deep-sea dumping of the Spar would then look very much like a precedent in the Greenpeace sense.

Before finally leaving this ticklish topic of precedent it is worth pointing out that it can work both ways. With the onshore disposal of the Spar now agreed, will this create an equally irrational precedent in the opposite direction? Greenpeace is unlikely to object if it does.

Chapter 5

The siege

DAY 1: 30 APRIL 1995

On Sunday 30 April, the press newswires crackled with the story that a number of environmental activists from Britain, Germany and Holland had, using just wires and ropes, scaled and occupied an offshore oil platform 118 miles northeast of Lerwick, off the British Shetland Isles. The details were sketchy but it became clear that the activists were from the pressure group Greenpeace International and, it was reported, they had taken over the installation in an effort to prevent it being dumped in the ocean by its owners Shell Expro. By 6 p.m. that evening, both television and radio news programmes were covering the dramatic scenes of the 'North Sea platform siege'.

Over the next six weeks, information on many of the topics covered elsewhere in this book dribbled into the public domain through the media and from the PR departments of the principal protagonists. This chapter attempts to chronicle the events as they were reported, without correcting any of the resulting confusion and misinformation. This garbled story, often sidetracked by irrelevancies and trivia, would have been the way in which most of the European public learned about the Brent Spar.

Greenpeace had enlisted four experienced alpine climbers to carry out the initial scaling of the Spar's exterior. They were then followed by six activists and tons of supplies and equipment. The operation had been organised from a small rented office in Lerwick and was instigated after Greenpeace received word of Shell's intention to dump the Spar in the deep ocean. A suitable weather window was identified around 30 April and so Greenpeace chartered the Danish vessel *Embla*, filled it with the mountaineers and supplies, left

Hamburg on 28 April, and met up with the Greenpeace ship *Moby Dick* to the north of the Spar on 30 April.

A Greenpeace statement at the time declared 'Dumping oil platforms laden with toxic chemicals will only add to the list of contaminants and proves yet again that the UK Government and the oil and gas industries have total disregard for the health of our seas'.

The group also maintained that the Brent Spar contained over 100 tonnes of toxic sludge, including oil, chromium and lead, as well as more than 30 tonnes of 'radioactive material' such as cadmium and arsenic. The activists claimed they would occupy the platform until the decision to dump the installation was reversed. They also claimed to have taken on board enough food and supplies, along with a generator and communications equipment, to last them for 'what is expected to be a long occupation' (see figures 5 and 6).

DAYS 2 AND 3

On 1 May Shell Expro confirmed that 'several persons' had boarded the Spar without permission and that they had alerted the local police to their presence. The same day a force-nine gale battered the area surrounding the Spar.

By 2 May officers from the Grampian police, the police force with jurisdiction for the North Sea oil industry, had been flown offshore to the Brent field complex, approximately two miles from the Spar, to begin an investigation of Greenpeace's occupation. At the time it was considered unlikely that the police would try to evict the protesters. It was thought, instead, that Shell would bring a case alleging trespass against Greenpeace and sequester its assets, thus forcing its activists off the platform. The police confirmed, however, that they had held emergency talks with Shell senior management at which 'all options of ending the sit-in' were discussed.

Up to twelve activists were reported to be occupying the Spar by this time. Greenpeace was now claiming the plans to dump the installation contravened the 1972 London Convention, which outlawed the dumping of man-made objects in the sea, and the 1992 Oslo-Paris Convention, which proscribed disposal in the sea of anything which might harm marine life.

At this stage, Shell did not dispute Greenpeace's claim over the amount of residual toxic waste left on board the Spar; it was, after all, mainly based on Shell's own figures. But Shell claimed that the ocean currents would quickly and harmlessly dilute the quantity. Shell also confirmed that the waste from the Spar would have an impact on marine life within a few hundred metres of the proposed dumping site – however, at this stage the precise location had not been made public, although a shortlist of three sites was being considered.

Shell also defended its position regarding the effect of the radioactive materials on the Spar. It maintained that the radioactivity came from salts that are a normal part of the marine environment and that 'the total radioactivity in the Spar is roughly equivalent to a medium-sized street of granite buildings [in Aberdeen]'.

Another reason put forward by Greenpeace to explain its actions was its claim that if the Spar dumping went ahead, it would set a precedent for the four hundred or so other North Sea platforms earmarked for disposal over the next few decades, as the North Sea oil boom slowly diminished.

DAY 4

On Wednesday 3 May, Greenpeace stocked up with supplies. 'We've been busy today loading food and equipment on board. We are here for a long stay', stated campaigner Tim Birch via a satellite telephone link. The protesters were adamant they were going to stay on board the Spar until the autumn when the weather would become so bad that any plans to sink the Spar would have to be postponed until the following spring or summer. Seven journalists from Germany, the Netherlands and Britain joined the protesters at this time.

On the same day, the seven detectives investigating the occupation from the Brent Alpha field flew back to Aberdeen to prepare a report to present to the procurator fiscal. Customs officers also boarded the Greenpeace supply ship the *Moby Dick* in an attempt to obtain a crew listing. This would inform the investigating officers exactly who was involved with the occupation. A police spokesperson also denied rumours that the SAS were going to be called in to remove the protesters. Greenpeace claimed that if the SAS were used it would refuse to be drawn into any violent confrontation with them. Sue Cooper, a Greenpeace spokesperson, at the time said 'we are a non-violent organisation, we are not threatening anyone with violence and we will not resist with violence because it is strictly against our code'.

Meanwhile, the inevitable storm was beginning to rage in the world's press regarding the possible adverse effects of dumping the Spar, and possibly other installations, in the sea. These disposal plans would amount to the biggest, costliest waste management exercise that the UK had ever undertaken. Andrew Searle, spokesperson for UKOOA, insisted that three-quarters of all UK sector installations (over 200) would be floated, towed away and scrapped onshore. But the remaining quarter – approximately fifty, mainly gigantic concrete and steel structures, could, it was argued, be left in place or toppled *in situ*. These two options provided the least risk to the safety of the men carrying out the work. However, this statement was not saying anything new. The Government had already ruled out simple abandonment of any of them, as the platforms would be a serious hazard to shipping and marine life as well as infringing a number of international treaties.

If the Spar itself was considered, onshore scrapping was thought to require around 360,000 man-hours of work. In addition to this, the complexity and labour-intensive nature of this option would also greatly increase the cost of decommissioning and expose personnel to greater risk. In contrast, deep-sea disposal was estimated to require only approximately 52,000 man-hours of work and involve less dangerous operations. Another possibly important point was that the costs involved in bringing the Spar ashore would be in the region of £46 million, and to dump at a deep-sea site, approximately £12 million.

In the general case for decommissioning, each platform owner would choose which option it wished to adopt. The choice of disposal method is an important decision, as up to 70 per cent of disposal costs are deductible from the Petroleum Revenue Tax. UKOOA estimated that its members would spend £1.5 billion on decommissioning structures in 126 oil and gas fields over the next ten years. An analysis for the newspaper *Scotland on Sunday* by oil consultants Wood Mackenzie, estimated that the total costs of 'abandoning' existing, planned-for and potential production platforms and other installations over the next thirty-five years would be in the region of £9.2 billion (in 1995 monetary terms). This figure did not include the cost of clean-up or the removal of thousands of miles of sub-sea pipelines.

Another issue that began to emerge was the long-term residual liability of the owner of the installation after decommissioning has been completed. Companies maintain that once they have fulfilled the criteria for decommissioning a particular platform, they should no longer be liable for any incidents involving the structure, or what remains of the structure. Governments should then take over responsibility, they argued, with the industry paying a type of premium insurance cover to assist with any such claims.

By now further details of the dumping site were emerging. It was said that the site earmarked for disposal was over 2,000m in depth and about 250km from land. It was also claimed, by Shell, that there is little marine life on the seabed at such a depth and thus the environmental consequences of dumping would be negligible. Greenpeace, on the other hand, claimed that dumping the Spar would release organochlorines such as polychlorobiphenols (PCBs) which have been found to accumulate in marine mammals such as whales. Although it is a difficult task to positively assign the source of the accumulated chemicals, Greenpeace reported the example of a sperm whale found on a Belgian beach which had such high levels of chemicals in its blubber that it was declared by the authorities to be 'toxic waste'.

Greenpeace also had support from the Wildlife and Countryside Link's Oil and Gas Consortium. Dr Sian Pullen, a marine conservation officer for the World Wide Fund for Nature UK, wrote to the Scottish Secretary on behalf of the Consortium pleading for caution in allowing the dumping to go ahead.

> We believe that this case must not be considered in isolation, that there needs to be a strategy for the decommissioning of all existing oil and gas installations and that before any new development is considered, a full 'cradle-to-grave' analysis of all aspects of the development, including the decommissioning of installations, must be undertaken. Indeed, a decision to dump this structure in this manner will set a precedent for the future with unfortunate consequences for the marine environment.

Dr Pullen added that Shell's decision had been reached using 'inadequate consideration of all possible options' and she questioned their ability to select the Best Practicable Environmental Option using 'inadequate technical information'. She went on to say that research carried out on the eastern seaboard of the United States of America had found that the highest diversity of marine life was to be found between 1,000m and 2,000m depth of water. Hence it was unwise to claim that there was minimal marine life at depths of 2,000m or more. The deep waters are still, to a large extent, an unknown quantity.

DAY 5

By Thursday 4 May, the protesters were settled in and making themselves at home in their unusual habitat. They had a French chef on hand to prepare their food and they were regularly being supplied with drinking water, food and replacement crew members by ship. The Spar had become Greenpeace's unofficial North Sea office with communications equipment, a fax machine and satellite telephone. They were living in the uppermost accommodation deck nearly a hundred feet above sea level. Seawater to flush the only toilet available to them was hauled up in a bucket.

Thursday also saw the publication by the DTI of draft guidelines for a protocol to be followed in the decommissioning, including the partial or total abandonment, of offshore structures. This document re-emphasised the already well publicised commitment of the UK Government to the case-by-case approach to disposal, apparently to counter the Greenpeace claim that the Brent Spar disposal plan would create a precedent. But the guidelines were widely interpreted as supportive of the offshore oil and gas industry if they wished to leave obsolete installations at least partially in place rather than removing them. One decision, to reduce the depth of clear water to be left over any remnants from 75m to 55m, particularly incensed the North Sea fishermen whose spokespersons, only a couple of days previously, had broadly been supportive of the Spar disposal plan.

DAYS 6 AND 7

Friday 5 May saw the beginnings of a story that must have contributed significantly to the way many members of the public viewed Shell and the oil industry in general, that of the infamous 'concrete eggs'; this tale is recounted in Chapter 1 and so will not be repeated here.

Friday was also significant in that it was the date that the all-important deep-sea disposal licence was finally issued to Shell Expro by the DTI. It was a common misapprehension by the general press that this licence had already been granted in February, but this was not the case.

At the end of the first week of occupation, Shell and the Grampian police warned Greenpeace that there was a possible danger of toxic and explosive fumes and warned the activists not to enter any enclosed areas. Greenpeace retaliated by saying it had taken the relevant precautions. Its team had breathing equipment, fire extinguishers and fume-sensing devices.

It soon became clear that the timing of the occupation was no accident. Greenpeace believed that the publicity resulting from its occupation would greatly enhance its cause at the ministerial conference to be held in Denmark in June. This meeting was to concentrate on developing measures to protect the North Sea from pollution and over-fishing.

At this time Greenpeace was basing its case on the expert reports, written by Rudall Blanchard, which were commissioned and made public by Shell, and their own report, written by an independent offshore engineering consultant, Mike Corcoran, which, not surprisingly, questioned Shell's justification of its chosen option.

DAY 12

On the 11th, the Scottish Office issued the final permission required to dispose of the Spar in Scottish waters. On Friday the 12th, two court officers flew to the rig to serve the activists with legal documents, granted by an Edinburgh court, ordering them to leave the Spar. In a 'bizarre twist', as reported by the *Evening Express*, the helicopter arrived so late on the Spar that the officers had to spend the night on the installation and were treated to breakfast by Greenpeace before leaving in the morning. The interim injunction required the activists to leave the platform immediately and not to impede any future action necessary to dispose of the installation. At the time, however, the court officers had no instructions to take any steps to remove the protesters physically, but were empowered to obtain their names and addresses. In a counter-move, Greenpeace considered suing the Government for granting Shell permission to bury the Spar at sea.

In Europe, support for the protesters was growing. Ritt Bjerregaard, EU Commissioner for the Environment, indicated on

Danish television that she would press for an EU ban on dumping at the North Sea Conference of Environment Ministers due to take place in June, if the ministers involved failed to address the matter themselves. Quoted in a Greenpeace press release on 13 May, she commented that 'most countries in Europe think [the dumping of the Brent Spar] is dirty and should be stopped', and went on to say that 'it is good that Greenpeace is around to ensure these things do not go on secretly'.

Greenpeace, unsurprisingly, failed to respond to the order to quit the Spar. The activists were by then in breach of the interim interdict and a Shell spokesperson said that the Company would 'fairly swiftly' refer the matter back to the Court of Session for further consideration. The UK police had decided earlier not to enter into criminal proceedings against the group, leaving Shell to pursue matters through the courts.

As Greenpeace held on to the Spar, the Labour party joined the growing barrage of criticism of the disposal plan. The Opposition party claimed that the Government-authorised plan to dispose of the Spar compromised accepted environmental standards for short-term economic benefit. Frank Dobson, the then Shadow Environment Secretary, condemned the Government for giving its approval to dump the Spar. 'This dumping policy is bad for the marine environment and could be bad for navigation', he said. Shell refuted this claim and stated that an independently assessed report had highlighted dumping to be the best option from an environmental point of view and in terms of several other considerations including health, safety and economic efficiency. Shell also reported it had spent more than one million pounds on surveying alternative deep-water sites for sinking.

Meanwhile, Shell had asked the Court of Session and the presiding judge, Lord Johnston, for an order for summary ejection of the protesters. Shell was still unable to furnish the court with the names of the protesters, and the main objective of the judge was to decide whether or not the court could order unnamed persons to be evicted under Scottish law. Mr Gerald Moynihan, acting on behalf of Shell, informed the judge that Lord Kirkwood had already granted an interdict to ban the protesters from occupying the platform. Mr Moynihan informed the court that the protesters were trespassing and had refused to leave the Spar after the interdict had been served. Hence they were in contempt of a court order. Shell had a right under common law to remove them. He went on to say that the issue was whether or not the court would make an order of ejection and hence assist Shell in removal with a minimum risk of confrontation:

> If your lordship refuses, you will be signalling the impotence of this court against people who sit on an oil rig in the North Sea and, in effect, blow raspberries. Can the court, in the face of stubborn resistance from a class of people intent on resistance, do nothing? That would be an absurdity.

He went on to state that Shell did not want the protesters jailed. They were seen as martyrs to a cause and jailing them would simply fulfill their wish for ongoing publicity. He simply wanted them removed and transferred back to the mainland.

Mr Jonathan Mitchell QC, acting on behalf of Jonathan Castle, one of the protesters aboard the Spar and the only named member of the Greenpeace crew, argued against an eviction order against his client. Mr Castle had not been given the chance to take legal advice regarding the interim interdict and so it could not be said that he was in contempt of court at this time. He also argued that Shell's demand to evict the other eleven unnamed protesters was impotent, as it was a cardinal feature of Scottish law that court orders could only be made against named individuals. In addition, he claimed that the majority of the other protesters served with the interdict on the Friday previous had now left the Spar and had been replaced by other Greenpeace members, and hence could not be in contempt.

DAY 20

On 19 May, Lord Johnston ordered the one named protester, Jonathan Castle, to be ejected, but ruled he could do nothing about the other eleven thought to be on board, as he did not know their names. He was in little doubt that the protesters were fully aware that they were flouting the court order. However, the problem of granting an order against unnamed people was that, if the original members of the team were no longer on the rig, the order would not pertain to the people on which the original order had been served. Jonathan Castle was also ordered by Lord Johnston to reveal the names of his fellow protesters so that Shell UK could then obtain the relevant named orders against each one. He also added that it should be made very clear to Mr Castle that the court would take action if he refused to disclose the identities of the other people involved. This order allowed Messengers-at-Arms to board the Spar and forcibly evict any named protesters, if necessary.

Shell's failure to obtain an order that covered the whole group of protesters must have been frustrating for the company. There now seemed to be a large amount of confusion over the eviction order given by Lord Johnston. Did the order imply that, should an eviction take place, action would be taken solely against Castle or against the whole group? Shell would not comment on what action it now wanted to take but was said to be liaising with the sheriff officers who were now responsible for carrying out the order. Police would be present when the eviction occurred, to ensure safety and make sure no laws were broken, but would not involve themselves further in the proceedings.

Meanwhile, in London, a judge adjourned a Greenpeace application for permission to bring a High Court challenge over the Spar. Greenpeace was applying for a judicial review of the Scottish Secretary's decision to grant the licence for disposal. Mr Justice Popplewell gave Greenpeace's lawyers until Wednesday 24 May to produce evidence to support their contention that the Scottish courts were unlikely to recognise the environmental pressure group and this amounted to 'special reason' for a hearing going ahead in London – as England, they alleged, would be affected by the dumping licence.

On the continental mainland, the European Parliament passed a resolution urging a halt to the Brent Spar dumping, demanding the European Commission 'take all possible steps to prevent the dumping'.

DAYS 20 AND 21

Greenpeace began to fortify the installation over the weekend of May 20–21 against a possible siege by Shell. Metal bars were welded across windows and the helicopter deck was rendered unusable with barricades and netting placed across it. More frivolously, party balloons and strings of banners were attached to the heli-deck to help prevent helicopters landing. The weather seemed to be in the protesters' favour, with wind and rain lashing the rig. The one named member of the team, Jonathan Castle, a former captain of the Greenpeace flagship *Rainbow Warrior*, was in the most danger as he now carried a warrant against him. He prepared to hide away in a barricaded room should Shell manage to gain access to the platform.

On Saturday morning, the police called to ask if they would be granted permission to board the Spar to ensure 'that justice is done', but were refused access.

On Sunday the 21st, at 11 a.m., a helicopter was spotted. It came near to the Spar, hovered and circled before flying off again. The protesters believed this to be a reconnaissance mission before a main attack; it helped to convince them that an attack was imminent. The court order had been granted on the Friday, and Saturday the 20th had been the starting date for the contractors to begin removing the remaining equipment on the Spar before it began its final journey to be dumped. Also on the weekend, the *Stadive*, Shell's gigantic, self-propelling, multipurpose maintenance vessel, was spotted approximately twenty miles away from the Spar. The vessel dwarfed the Spar in height and contained cranes that could simply drop boarding parties onto it. Shell refused to be drawn on what the rig was doing in the neighbouring waters, except for the fact that it was carrying out normal maintenance work in the Dunlin field about fifty miles away. But Shell did not rule out that it would be used for boarding officials and company men. By Sunday night, the *Stadive* was manoeuvering itself alongside the Spar, using its computer-controlled positioning system to help keep it stationary without the need for anchors.

DAY 23

In the early hours of Monday 22 May Greenpeace activists chained themselves to equipment on the Spar as an attempt was made by Shell to reclaim the platform. Shell, under the cover of darkness, had used a crane, mounted on the *Stadive*, to try to winch a container full of men aboard the ten-storey high Spar. It failed in its first attempt to get the container onto the heli-deck, and deteriorating weather conditions, with force seven winds and high seas, forced Shell to abandon the attempt. When dawn broke, two Shell vessels were spotted alongside the Spar; however, Shell claimed, the forecasts of worsening weather had postponed the operation.

DAY 24

On 23 May, Shell announced that it had taken back possession of the Spar after twenty-four days of occupation by Greenpeace. Police and Company officials staged another dawn raid at 6 p.m. on Tuesday morning using the *Stadive*'s crane to hoist a steel cage of men onto the platform (see figures 7 and 8). The activists vainly attempt- ted to push the container away from the deck. Inside the container were sheriff's officers, four policemen, Shell security staff and Diarmid McAlister-Hall, the Brent Spar manager. Under the 1971 Offshore Workings Act, Mr McAlister-Hall had total control over any personnel aboard his installation. He read out a statement explaining he had the eviction order and would detain people on the structure.

'The process of removing the illegal occupiers continues in a safe and controlled manner', said a statement from Shell. Bolt-cutters were used to release protesters who had chained themselves to the platform itself. Greenpeace claimed, at the time, that five of its protesters were arrested, and that a number of them, including Jonathan Castle, were in hiding. By Tuesday afternoon, around twenty protesters and journalists had been removed. This left two campaigners on the upper levels of the structure and Castle secreted somewhere below decks like a sort of latter-day Pimpernel or a forerunner of the fossorial 'Swampy' who later achieved a similar short-term notoriety during the campaign against the Newbury bypass. It was rather ironic that, of all the protesters, Mr Castle was the only person whom the sheriff's officers were empowered to remove from the Spar, and he was one of the last to be still at large. Shell continued with its level-by-level search on the structure using breathing apparatus, as Greenpeace defiantly vowed to continue its campaign.

Fourteen hours of hunting by police and Shell workers eventually unearthed Jonathan Castle, who was hiding in the depths of the Spar's interior. The searchers braved flares and smoke canisters set off by protesters. They eventually found Castle in a welded compartment where he had barricaded himself. As he was winched off the Spar that had been his home for three weeks, he cheerfully waved to his two colleagues still perched in home-made towers above the Spar deck, but later ordered them to give themselves up. Shell did not seem unduly worried about the two remaining protesters perching in their bird's nest vantage points: 'We are not concerned about them. Our aim is to decommission the rig and we can now proceed with that', a Shell spokesperson was quoted as saying.

On the same day, Greenpeace activists in The Hague blocked entrances to the Dutch Headquarters of the Royal Dutch/Shell group and hung banners from the main gates of the building which read 'stop dumping platforms'. They also formed a human chain around the building to protest about the company's action against their colleagues on the Spar.

DAY 25

On the 24th, one day after Greenpeace lost control of the Spar, they were in the High Court seeking leave before Justice Popplewell for a judicial review of the Government's decision to grant Shell permission to dump the Spar. The group's argument centred around its belief that the Government's approval breached three international agreements, drawn up to avoid marine pollution through man-made sources. Edward Fitzgerald, Greenpeace's lawyer, said that disposing of the Spar in the Atlantic would be in breach of the 1972 Oslo Convention on the prevention of marine pollution by dumping from ships and aircraft. The judge could have ordered a temporary revocation of Shell's disposal licence, forcing it to abandon its ten-week pre-burial schedule. In the event, Mr Justice Popplewell ruled that it was not in his jurisdiction to review 'the making of what is essentially a foreign administrative decision'. Greenpeace failed in this action as the court ruled that it was a Scottish case and that it would be improper for an English court to seek a review.

Though previous action had taken place in Scotland, Greenpeace had attempted this legal action in England because it did not have legal status north of the border. Greenpeace's legal director, Sarah Burton was quoted at the time: 'Our advice was that there was jurisdiction in either court because the Secretary of State for Scotland has an office in both Scotland and England and the decision he made was a decision of the UK Government'.

She went on to add: 'The effect of that decision would be equally felt in every part of the UK because it involves an installation at sea'. In a press release Greenpeace claimed that the decision had hinged on a 'legal technicality' and that the decision meant that no British court could hear a Greenpeace challenge to Government. 'In effect, it was a decision by the UK Government to allow Shell to dump toxic chemicals and radioactive waste at sea'.

More news of the proposed dump-site was also leaking out. It was thought to be one of three deep-sea sites in the North Atlantic: the North Feni Ridge, the Rockall Trough or the Maury Channel. It emerged that the final choice was the Feni Ridge site, located 150 miles northwest of Lerwick.

At the end of May, the German government appeared to change its view over the dumping of the Spar and sided, albeit cautiously, with the British Government. Angela Merkel, Germany's environment minister, had originally opposed the move to dispose of the Spar, along with the European Parliament, and had lodged a protest with the DoE. However, after a meeting with John Gummer, the then British Environment Minister, she was convinced that the Government had correctly decided that sinking was the best option in this particular case. 'Our discussions have convinced me that Britain's policy will be to use these disposal methods only in exceptional cases', she told a press conference in Bonn.

At the same briefing John Gummer informed the assembled press that the disposal of the Spar would be carried out in full accordance with international agreements and would not harm the environment. 'The best option is dismantling at sea. Shell are required, of course, to remove the heavy metals and anything that would cause damage'.

This cautious alliance between Germany and Britain regarding the Spar came at a very fortuitous time for the UK, as the fifth North Sea Conference was about to begin at Esjberg, Denmark. Britain was steeling itself for heavy criticism from governments such as the Belgian, Danish and Icelandic, who had already expressed alarm that the Spar's disposal would set a precedent. But with the Spar back firmly under Shell control, the deep-sea disposal seemed destined to go ahead.

Figure I The Brent Spar operating in the Brent oil field in the North Sea, off-loading oil into a tanker for transport ashore.

Source: Shell UK Exploration and Production

Figure 2 The Brent Spar moored in Erjford, Norway, after Shell's capitulation, but still floating as she had during her working life. The height from the sea surface to the helicopter deck was 28 metres, with the 26 metre diameter superstructure standing on a 17 metre diameter column disappearing beneath the water.

Source: Shell UK Exploration and Production

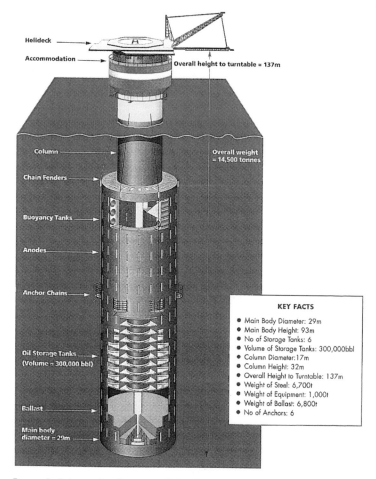

Helideck

Accommodation

Overall height to turntable = 137m

Column

Chain Fenders

Buoyancy Tanks

Anodes

Anchor Chains

Oil Storage Tanks
(Volume = 300,000 bbl)

Ballast

Main body
diameter = 29m

Overall weight
= 14,500 tonnes

KEY FACTS

- Main Body Diameter: 29m
- Main Body Height: 93m
- No of Storage Tanks: 6
- Volume of Storage Tanks: 300,000bbl
- Column Diameter:17m
- Column Height: 32m
- Overall Height to Turntable: 137m
- Weight of Steel: 6,700t
- Weight of Equipment: 1,000t
- Weight of Ballast: 6,800t
- No of Anchors: 6

Figure 3 Schematic diagram of the Brent Spar showing the relationship between the familiar visible section and the submerged main bulk of the structure. It also shows the compartmentalised internal construction of the oil storage tanks.

Source: Shell UK Exploration and Production

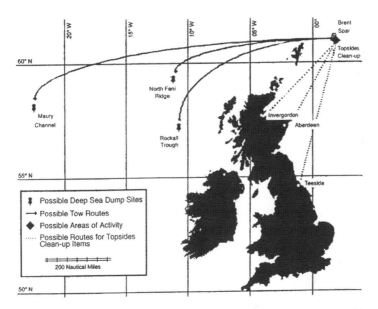

Figure 4 Map showing the working position of the Spar in the Brent
Field and the routes to the three possible deep-sea disposal
sites and to the UK onshore facilities where the topsides
might have been dealt with. The North Feni Ridge site, about
150 nautical miles to the northwest of the Hebrides, was the
one ultimately selected and for which the Government
disposal license was issued.

Source: Shell UK Exploration and Production

Figure 5 Greenpeace climbers scaling one of the Brent Spar's anchor chains to fix a banner to the side of the Spar's column.

Source: Greenpeace

Figure 6 Pictures of the Greenpeace-occupied Brent Spar were flashed around the world in early May 1995, dominating television news broadcasts and the front pages of newspapers. The 'SAVE THE NORTH SEA' banner, ostensibly to draw attention to the upcoming conference of North Sea environment ministers, gave many onlookers the erroneous impression that this is where Shell intended to sink it.

Source: Greenpeace

Figure 7 Tuesday, May 23 1995; the re-taking of the Spar. Sherriff officers, police and Shell personnel are lifted in a cage from the
support vessel *Stadive*, while Greenpeace activists aboard the
Spar try to prevent it being landed on the helicopter deck.

Source: Greenpeace

Figure 8 The re-taking of the Spar. Despite the barricades of poles, wire and netting erected by the Greenpeace occupiers, the Shell personnel and police get on the helicopter deck. Now they have the problem of flushing out the hidden protesters.

Source: Greenpeace

Figure 9 David and Goliath; the essence of the affair in the eyes of many observers. Greenpeace activists in a vulnerable little rubber boat overshadowed by oil giant's fortress-like Brent Spar.

Source: Greenpeace

Figure 10 The second Greenpeace occupation of the Spar, in mid-June 1995, while the structure was being towed towards the proposed deep-sea disposal site. By this time the conference of North Sea environment ministers was over and the Spar was well outside the North Sea. The 'SAVE THE NORTH SEA' banner had been changed for one simply saying 'STOP SHELL NOW'

Source: Shell UK Exploration and Production

The hull is cut into sections and skidded on to a barge for transportation

Spar is raised vertically using jacked cables and a lifting cradle

Spar with lifting barges

Concrete quay is laid on top of Spar "slices"

Spar "slices" are positioned for quay extension

View of existing quayside at Mekjarvik

Figure 11 The existing quayside at Mekjarvik, Norway, before the installation of the proposed extension using 'slices' cut from the Brent Spar.

Source: Shell UK Exploration and Production

Figure 12 Brent Spar's final resting place: an artist's impression of the completed quay extension and ferry terminal at Mekjarvik, Norway.

Source: Shell UK Exploration and Production

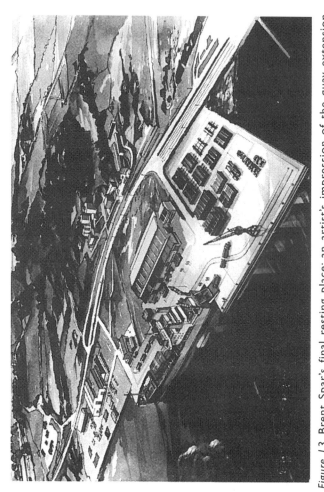

Figure 13 Brent Spar's final resting place: an artist's impression of the quay extension and ferry terminal at Mekjarvik

Source: Shell UK Exploration and Production

Chapter 6

The capitulation

In June, media interest in the Spar reached its zenith. The drip-feed of new information regarding its alleged state of contamination, and the possible consequences of dumping, continued to leak out in a confusing and sometimes contradictory manner. As with the preceding chapter, here no attempt is made to filter out any misinformation. Instead events are chronicled as they appeared in the press. Any remark by the authors, to help explain or comment on the reporting, is set in italics.

Following Shell's recapture of the Spar, the company began defending its decision to sink it. It claimed that the Spar's draught, at 109m, made towing it in for disposal on British soil impossible, as the sea is less than 85m deep around the mainland coast. The only feasible way of bringing the Spar inland was to turn it on its side, a dangerous operation that could result in the Spar sinking, with possible loss of life to the personnel involved. Shell also launched a propaganda counter-offensive against Greenpeace, reiterating the argument that the decision to dump came after three years of in-depth analysis of disposal options and that this included: 'the completion of 15 separate studies on the environmental and engineering aspects, taking independent surveys and advice into account, and the careful weighing of all environmental, safety, public health and economic considerations'.

Also, in a burst of generosity, the company dropped its legal action against the Greenpeace protesters; a spokesperson announced that Shell would not continue with the court order forcing Greenpeace to name all of the Spar's illegal occupiers.

On Tuesday 5 June, Greenpeace protesters boarded the tugboat

Smit Singapore, one of the vessels being used to tow the Spar. They chained themselves to mooring ropes whilst divers attempted to immobilise the propellers. The protesters were eventually removed, but Greenpeace claimed success in that the occupation had delayed the vessel's departure by twelve hours.

On Wednesday the 6th, Greenpeace activists once again occupied the Spar. They climbed up an anchor chain which gave them access to a closed-off, lower-level walkway. Five members were reported to have spent a few hours on the installation before rejoining the *Moby Dick*. The reboarding was made possible by Greenpeace's diversionary tactics back at Aberdeen harbour, which kept police and Shell officials distracted. Shell warned that the Spar had been fitted with explosives in preparation for dumping. Greenpeace claimed that Shell employees used water cannon against them, one woman being allegedly fired upon for a whole twenty minutes. Shell denied these claims, but did suggest that the occupiers were putting themselves at risk by reboarding. It maintained that this further action would not affect the company's plans and warned that preparations for disposal were nearing completion and as such, the platform was becoming increasingly dangerous to unofficial visitors.

7–14 JUNE

Greenpeace's activities coincided very neatly with the fourth North Sea Conference in Esbjerg, which started the following day, Thursday 7 June. The UK had made major concessions at the previous conference in 1990. Facing widespread charges that the UK was 'the dirty man of Europe', the then Environment Secretary, Chris Patten, was forced into expensive compromises shortly before the Conference began. The Government had been hoping for an easier ride at the fourth Conference; Greenpeace's occupation of the Spar had effectively scuppered its plans.

The Conference began with a bang, with ministers from five countries – Sweden, Denmark, Belgium, Germany and the Netherlands – launching a sustained two-pronged attack on Britain for granting permission for the Brent Spar to be dumped at sea and for allowing excess nutrients from sewage to enter the North Sea.[1] They had ostensibly gathered to finalise agreements to help clean up the North Sea and protect its dwindling fishing stocks, but these two issues were the sticking points, Britain being the key 'villain' on both counts.

The Swedish environment minister, Ms Anna Lindh, accused Britain of 'greed and a lack of foresight' on the Spar issue. She was joined by Ritt Bjerregaard, the EU Commissioner for the Environment, who maintained that the total cost to the ecosystem of the area should be considered, not just the short-term cost advantage of dumping. She went on to say that: 'The dumping of the Brent Spar is not acceptable. The dumping of all old oil installations must be banned'. She went on to call it an 'act of environmental vandalism' which future generations would find hard to forgive.

John Gummer immediately went on the offensive. He tried to take the focus away from the Spar by accusing the Danes of 'appalling industrial fishing practices'. He added that Britain would not be bullied over the Spar's fate, even if it meant putting the success of the Conference in jeopardy.

The chairman of the Conference, Svend Auken, the Danish environment minister, stated in his opening comments that 'There should be no more free rides. Not all states have fully understood this'. Speaking specifically about Britain's failure to remove nitrate and phosphates from sewage before it is pumped into the sea, he went on to say 'We have not succeeded as North Sea states if any of us continue to use the seas as a rubbish dump'. Although the Brent Spar issue was not directly related, its presence exacerbated the already fraught relationship between the UK and other North Sea countries.

On the nutrient issue, Gummer believed that the algal blooms, a result of excess nutrient concentration, were not a problem on the British side of the North Sea and hence no action needed to be taken. The continental countries, on the other hand, wanted the North Sea to be declared a sensitive area and a vulnerable zone. This would result in phosphates and nitrates having to be extracted from sewage before disposal. It was estimated that the cost to east coast water companies of cleaning up the sewage could be up to £2 billion.

As far as the Spar issue was concerned, Britain had only one strong ally. Norway, with potential dumping issues of her own, sided with the UK. Although, to date, Norway had always decided on land disposal, it was thought that a question mark might hang over future decommissioning plans. Siding with Britain, it called for a case-by-case basis for the decommissioning of oil platforms.

The stance of the five European ministers did attract some criticism, however. It was suggested that the countries lambasting Britain were only too delighted to have the attention of the meeting focused on decommissioning issues, presently concerning only Britain and Norway, rather than the potentially more environmentally damaging issue of over-fishing which affects all countries bordering the North Sea.

On Friday, the final day of the Conference, the environment ministers of all the North Sea countries, except for Britain and Norway, supported a resolution calling for all offshore installations to be dismantled on land.

Back in the North Sea, Greenpeace again clashed with Shell as the company's workers were raising the installation's anchor chains to allow it to be towed. The protesters, in inflatable dinghies, fixed themselves to the chains in a bid to stop Shell raising them. Greenpeace claimed that three protesters were thrown into the sea after being rammed by Shell's inflatables and another had fallen in when Shell tried to winch on board a Greenpeace boat, flipping it over in the process. Shell retorted that it had, in fact, rescued some of the protesters on a Greenpeace life-raft and categorically denied any claim that it had been firing water jets at them.

Greenpeace had also made public a two-hundred-page unpublished Shell document. The group maintained that this showed that the platform could have been dismantled on land for less than a quarter of the cost claimed by the company. The report, 'Feasibility study for the scrapping of the Brent Spar', was written for Shell by Smit Engineering in 1992. It suggested that towing the platform to a Norwegian Fjord and scrapping it there would cost less than £10 million, while Shell's official publicly available estimate of scrapping onshore was £45 million (of which the British taxpayer would have to pay approximately £34 million in tax rebates). Smit also reported that there were no technical difficulties bringing the Spar to shore in Norway and recycling the materials. The irony being that Smit was the company contracted to sink the Spar three years later.

Shell maintained that it had not presented this document to the Government when it was making its case because it was an old report, one of many, was 'not really relevant' and was only used as 'primary material for a full and thorough engineering and viability study'. It did admit that the £10 million figure was accurate but that the sum was only the demolition cost and did not take into account other factors, such as the cost to repair the damaged storage tanks. Greenpeace argued that the estimates in the Smit study seemed to cover the same ground as the later study by Rudall Blanchard, the report that had been presented to the Government.

By 12 June, the Spar was being towed by five vessels. Shell had already warned Greenpeace that any move to disrupt the operation could have 'severe consequences' and that interference could lead to the grounding of the Spar at an unplanned shallow-water location, which could be inherently more environmentally sensitive than the chosen dumping ground. Unmoved by this warning, Greenpeace chartered a yacht, the Dutch-registered *Altair*, and joined the *Moby Dick* at the Spar to monitor the tugs' journey. They maintained that their activists would attempt to disrupt the sinking procedure once the Spar reached the North Feni Ridge. The journey was estimated to take one week.

On the 13th, Lord Peter Melchett, UK Director of Greenpeace, was ordered to appear before a Scottish judge and reveal the identities of the activists threatening to disrupt the dumping operation. Shell, it seems, had undergone another change of heart after abandoning its court action once it had retaken possession of the Spar in May. Jonathan Castle, now captaining the *Altair*, was also ordered to appear on the same day. On issuing the order, the Judge said: 'In my opinion, what exists here is an apparent blatant attempt to defy the authority of this court, which must be more important than any other consideration or cause, however genuinely felt or fought for'. Castle would have to explain his failure to obey the earlier court order that he reveal the identities of his fellow Spar 'squatters'. He would also have to explain the reports that Greenpeace intended to use the *Altair* to disrupt the sinking of the Spar.

Jonathan Castle, however, was more concerned about the safety of the towing procedure than getting to court on time. Greenpeace suggested that the towing could be in jeopardy after three buoys broke off from chains attached to the Spar. Shell denied there was a problem. Although they did admit that two of the buoys had broken free, they maintained that the actual chains were still in place and could be used if necessary. The Spar was being towed by two, out of a possible six, chains that had originally held it in place in the Brent field. The remaining four chains were connected to buoys and floated alongside the Spar as it was towed.

In the meantime, Shell was desperately trying to serve the latest court order on Mr Castle. Shell officials and sheriff officers had attempted three times to serve the papers on the elusive captain. Their first attempt involved abseiling under a helicopter over the *Altair* and reading the papers out to the occupants of the boat through a loudhailer, but this proved unsuccessful. Their second attempt involved using an inflatable to get alongside the *Altair*; a package was then thrown on board. Unfortunately for the sheriff officers, the package was 'inadvertently' swept back into the sea as the Greenpeace crew were busily washing the decks at the time. The third method involved attempting to fax Mr Castle; this also failed. Mr Castle was available to come to the phone, however, and said: 'My number one duty is to remain out here to stop the Brent Spar being dumped. I have a job to do and that job is to prevent further pollution of the sea'.

An amusing aside emerged in the diary column of the *Guardian* on 14 June. The story alleged that, after Shell had removed the protesters from the Spar, it returned all confiscated property to the group. The returned property included a Macintosh laptop computer which had three new messages added to its memory. The first read: 'To all our Greenpeace readers, we enjoyed crossing swords with you and hope to do so again. But please have a wash next time'. The second message was signed by: 'Rapists of the planet'. The third was by far the juiciest: 'To Captain Dickhead: You fucked up, you wanker. Don't mess with the big boys. Shell rules. Regards, Dick (business as usual) Parker'. Needless to say, Dick Parker, Production Director at Shell's Aberdeen office, denied authoring the message: 'sounds like Greenpeace rubbish as usual'. The only sticking point was that the dates of the messages coincided with the time that Shell had sole possession of the computer. 'We can't vouch for everyone', retorted Mr Parker. (*authors' note: Was he suggesting that Greenpeace had infiltrated the hallowed ranks of Shell headquarters itself?*)

15–20 JUNE

By mid-June, Shell was coming under increasing pressure to call off its dumping plan. Foreign diplomats joined with environmentalists to urge Shell to change its mind and Germany warned that it planned to raise the issue at the G7 summit, due to take place later that month in Canada. (*note: The German government seemed to have conveniently forgotten the concession its environment minister, Angela Merkel, had made after talks with John Major at the end of May – only two weeks previously.*)

Elsewhere, Shell was attracting criticism from unexpected areas. The Cleveland County Superannuation Fund, for example, which holds £16 million worth of Shell shares, put pressure on the company. Ian Jeffrey, chairman of the Fund, warned Shell that they should not ignore the damage that the continuing controversy could cause to their, and their shareholders', interests. In Germany, sales from Shell stations were hit heavily after pleas from environmentalists to boycott them. In Leipzig, local government barred its civil servants from signing official contracts with Shell. Michael Muller, the environment spokesman for the Social Democrats, suggested that the boycott should be extended to the company Esso which co-owned the Spar with Shell (*note: This being one of the first public mentions of Esso's role in the whole Spar saga.*)

Greenpeace stepped up its campaign to persuade British drivers to boycott Shell stations and pointed to the success of the campaign in Germany. Some German garages had, it was alleged, reported losses of up to 50 per cent, although Shell initially denied the claim. Deutsche Shell complained that the action in their country was unfair since it could not affect the actions of its sister company back in Britain.

On 15 June, Lord Johnston issued a warrant for Mr Castle after he failed to appear before the courts. The warrant allowed the Messenger-at-Arms to bring Castle before him to answer to the contempt of court charge. Lord Melchett, who could not appear because he was in Japan, sent a representative, Steven Thomson, Resources Director, to represent Greenpeace. He was asked to supply the names and addresses of the protesters aboard the Spar. He reported that all of Greenpeace's supporters were on a computer database and promised to check it for the protester's details and supply them to the court, if they could be found.

In an interview with the BBC's *Newsnight* current affairs programme that evening, the head of Shell's German operation, Peter Duncan, admitted that he was seeing 'very significant measurable losses of business' following Greenpeace's campaign: 'The short-term impact on business is as serious as anything that has happened in the recordable history of this company'.

Duncan went on to suggest that ineffectual and inadequate communication could be, in part, to blame for the unprecedented level of public and international outcry against the decision. 'It is quite clear that all those concerned in this decision have wildly underestimated the political implications', he claimed, and said that there had been 'an inability to communicate to people what is the logic of the decision [to dump]'.

Confusion also reigned within the ranks of Shell's presidency. In Rotterdam, base of the headquarters for the multinational parent company, Royal Dutch/Shell, a spokesperson, Henk Bonder, was quoted as hinting that Shell might change its mind. 'Our president Mr Slechte said that it must be a decision of the British Government to reconsider – if they do so, Shell would be the first one to talk with them'.

This statement clashed violently with the view of Shell UK. A spokesman for Shell Expro retorted: 'Mr Slechte is president of Shell Netherlands and he was not speaking on behalf of Royal Dutch/Shell....There is absolutely no question of us restarting negotiations or changing our position unless the government were to change its mind'.

As the G7 summit began in Halifax, Nova Scotia, the whole environmental agenda was given an unexpected push up the ratings by the announcement of the French government that it was to restart nuclear testing in the South Pacific. International outcry over this action established the environment as a main issue for the talks. The United States, Canada and Japan all joined to condemn the French stance and urged Mr Major to support their condemnation. The Prime Minister however, managed to avoid making an outright attack on the nuclear testing plans. In addition, Britain came under more and more pressure to withdraw its support of Shell's dumping plan. Chancellor Helmut Kohl and his finance minister Theo Waigel called for John Major to reconsider the Government's actions. 'My urgent advice is not to do it [dump the Spar]', said Mr Kohl.

By 16 June Shell was hotly denying rumours that it would postpone its plan to dump the Spar, presently forecast to reach its destination on 21/22 June. The rumour was fuelled by Shell's German sister subsidiary claiming that Shell would bow to public pressure and delay the procedure, this misinterpretation apparently arising from an inaccurate translation of a Shell statement sent to all its European subsidiary operations.

Shell and the British Government appeared to be standing firm against the mounting barrage of criticism both at home and abroad. The boycott of its products was gaining ground on the Continent, with German petrol stations being the worst hit. More seriously, in Frankfurt on Wednesday the 14th, six shots were fired at a Shell service station, damaging pumps and a display window. In Hamburg on the 16th, in the early hours of the morning, activists threw petrol bombs into a Shell sales room and spray painted the walls with the words: 'Don't sink the Brent Spar oil platform'. Greenpeace reacted quickly to dissociate itself from the violence, which was presumed to have been orchestrated by a radical German left-wing fringe group who regularly involve themselves with spats involving police and property.

The violent outbursts in Germany helped to persuade Chancellor Kohl to make a personal appeal to John Major. Mr Kohl discussed with Mr Major his troubles with the Green party in Germany and implored him to think again. The British Prime Minister, however, would not be moved and turned down the appeal. Mr Kohl said afterwards that they had discussed the issue but were unable to make progress. 'We talked about it and we were not in agreement. It's as simple as that', he said. He went on to say: 'We ask him to think again. We believe that it is not the looniness of a few Greens but a Europewide, worldwide trend for the protection of our seas'.

Greenpeace protesters occupied the platform once again on 16 June. Two protesters were dropped by helicopter onto the Spar's heli-deck; another was hosed off an inflatable whilst trying to clamber onto the structure. They dodged Shell's high-pressure water cannons to spend a cold and wet night clinging to the platform, despite the forecast of storm-force gales. With the *Altair* in pursuit of both the Spar and the two squatters, a Greenpeace spokesperson announced: 'Our two members are reported to be fit and well and willing to do everything possible to stop the dumping going ahead'. Greenpeace complained that Shell had used their water cannon to break windows on the Spar, a potentially dangerous activity. The water cannons were supposed to be creating a protective curtain around the Spar, but, Greenpeace maintained, they were being focused on the helicopter that was attempting to land the protesters. Shell strenuously denied these claims.

By Saturday the 17th, the same day as a Greenpeace-organised picket of Shell UK petrol stations, the pressure on the Government and Shell was reaching fever pitch. The papers were full of the embarrassment of Mr Major and his team in Halifax when, time and again, they had been attacked by their European G7 counterparts. Major had staunchly defended the deep-sea disposal decision, while in London Tim Eggar had tried to take the heat out of the situation by holding a press conference reiterating that all the many studies which had been carried out on the disposal options had demonstrated that the deep-sea one was the best.

Eggar also accused Germany of pandering to political pressures at home. Germany's green movement is much stronger than Britain's, and so the pressure to oppose the sinking would have been much greater there. He also criticised the quality of the German argument against sinking the Spar, and assigned their strong anti-dumping stance to 'domestic politics'.

Shell, on the other hand, did concede that the mounting criticism was 'highly unsatisfactory for an international enterprise with operations in all the countries concerned'. They promised also that they would make a better case to the public to persuade them that dumping was the most sensible option.

Greenpeace then threw more fuel onto an already inflammatory situation, by announcing publicly that they had estimated that the Spar contained more than 5,000 tonnes of oil, fifty times more than Shell claimed. They also added further details to the 'concrete eggs' story that they had alluded to earlier in the siege (see Chapter 1). (*note: Although both items were widely reported it seems unlikely that they made much difference to the level of public opposition to the deep-sea disposal, which was already intense. On the other hand, it is quite possible that they contributed to Shell's decision to halt the dumping. After all, Shell would have known that both stories were untrue. If the Spar was dumped, no-one would have been able to check the facts and the public would assume that the Greenpeace version had been correct. The company's only chance of restoring its reputation was to undertake an independent inquiry, ultimately the DNV assessment. For Greenpeace on the other hand, although it would have won in the short term, such an independent enquiry would be disastrous.*)

Further unfortunate timing for Shell included the presentation of its long-standing 'Shell Better Britain' environmental awards at a ceremony in London planned for Wednesday 21 June. The 25th anniversary celebrations of the 'Shell Better Britain Campaign' were, not surprisingly, postponed until a later date because of the escalating tensions. In a letter to more than 200 invited guests, which included representatives from English Nature and the Countryside Commission, Shell said that it had reluctantly decided to postpone the event because of the 'increasingly emotional debate'.[2]

The whole international debate on the Spar completely overshadowed what should have been another anniversary celebration, this time of North Sea oil's 20th birthday on Sunday 18 June. This being the date, twenty years previously, when the first oil from the sector came ashore at the Isle of Grain in Kent. In a statement marking the anniversary, Tim Eggar claimed that UK North Sea output was at record levels and rising: 'Recent announcements of major discoveries west of Shetland reinforce the view that there is still tremendous potential in the UK continental shelf, in areas that were thought to have little prospectivity only a few years ago'. He announced that UK crude output was at record levels of 2.6–2.7 million barrels per day (bpd). It was also reported that about 80 new oil and gas fields would be developed in the UK sector over the next 15 years, even if the rise in oil prices remained modest.

This anniversary heightened calls for Shell to be more environmentally responsible and recycle the Spar. The *Observer* reminded its readers that Shell had accrued a profit of £3 billion from the Brent field alone in the twenty years of North Sea oil. It argued that even if bringing the structure ashore did, in reality, cost £43 million, it would be a mere splash in the ocean in relation to Shell's profits. The article concluded by urging its readers to boycott Shell: 'Shell will only see sense if hit in the pocket. Join the boycott'.

Also on that Sunday, Chris Rose, Greenpeace's Campaigns Director, clashed with Tim Eggar on the *Breakfast with Frost* TV programme. The two came head-to-head in a heated debate that saw Eggar accusing Greenpeace of 'grossly exaggerating the problem, and completely misconstruing the independent studies that have been carried out'. The Energy Minister also claimed that 'The way Greenpeace has been handling this has put the lives of their own people at risk'. Chris Rose retaliated by accusing Shell of treating the Spar as if it were 'a gigantic piece of litter'.

Greenpeace were now saying that they would attempt to use their most powerful vessel, the *Solo*, ironically the former sister ship of the *Smit Singapore*, one of the tugs towing the Spar, to try and stop the dumping operation. The *Solo*, a converted North Sea 218ft tug, set off from Stornoway on Monday 19th, with a crew of seventeen, about twenty journalists from various countries and Councillor Iain MacSween from the Western Isles Islands Council, who was acting as an observer.

On Monday, Shell conceded that a combination of bad weather conditions and Greenpeace's sabotage attempts had delayed disposal. The operation was weather-dependent and, with two Greenpeace activists still hanging on to the structure, it was impossible to give a precise date for the sinking to take place.

By Tuesday, the flotilla of vessels surrounding the Spar was up to a total of nine. Greenpeace had the *Altair*, which was shadowing the Spar at a distance of two miles, and the *Solo*. Shell had six: the two tugs pulling the Spar, *Smit Singapore* and the *President Hubert*, and four supply ships. The ninth was the fisheries protection vessel *HMS Itchen*, there to supervise events on behalf of the DTI.

Meanwhile, on board the *Solo*, Greenpeace had told assembled journalists that the two activists on the Spar were prepared to chain themselves to the platform to stop the dumping. It was also confirmed that the helicopter pilot who dropped the activists on board had been arrested while refuelling on the Scottish mainland, and was helping the Scottish police with their enquiries; Uwe Lahrmann appeared in court on Tuesday the 20th, accused of reck- lessly and negligently endangering the German-registered helicopter and the people on board, as well as the boats in the flotilla of vessels that were surrounding the platform. The case was adjourned until 11 August and Lahrmann was granted bail on condition that he did not go within 500 metres of the Spar.

In The Hague, outside Royal Dutch/Shell's headquarters, Greenpeace dumped a 7m-high replica of the Spar, made from old oil barrels. Beside it they placed a portrait of the head of the company, Shell Chairman Cor Herkströter, with the message: 'If Herkströter has his way, this will be the only shell left in the North Sea'.

By this time, the Greenpeace assertion that the Spar contained much more oil and heavy metals than was admitted in the official Rudall Blanchard report, was attracting a great deal of interest. Greenpeace claimed it had taken the samples of the oily sludge whilst occupying the Spar in May (see Chapter 1). In a letter to the Shell UK board on 19 June, Lord Melchett reported a quantity of residual oil which was approximately fifty times greater than the concentration reported by Shell. Greenpeace also claimed that a quantity of toxic chemicals was discovered during its analysis of the sample.

On Tuesday 20 June, the day originally timetabled for the sinking, two more Greenpeace protesters joined their colleagues aboard the Spar in a dawn raid by helicopter. The helicopter had to avoid gale-force winds and the water cannon which were providing a protective curtain around the Spar. Greenpeace said that the four men hoped to hide within the structure to stop explosives being used to sink it (see figure 10).

On the same day Shell suffered a further setback when a memo came to light in which British scientists from the Ministry of Agriculture, Fisheries and Food (MAFF) claimed that waste from the structure would be a serious hazard to marine life. Shell retorted that this note had been written nearly two years previously, in December 1993, and referred to plans to dump in shallow waters – this plan was itself scuppered when deep-sea disposal was chosen as the best practicable option.

It was alleged that MAFF scientists were concerned about the threat to marine life if the Spar's oily sludge leaked out into surrounding waters. Their memo was itself leaked, to Greenpeace, and consequently distributed to the awaiting press. A handwritten note at the top of the document read: 'The bottom line is that the waste cannot be dumped at sea. The only option is to take it ashore and treat'. The handwriting was believed, at the time, to be that of Dr John Campbell, who was director of the Ministry's fisheries laboratory at Burnham-on-Crouch. The memo itself, being from one of his staff, read: 'The chemistry of this water is such that it has to be considered very toxic to marine biota life. It should therefore be treated as hazardous waste and discharge should be prohibited'. A MAFF spokesperson confirmed that the scientists had been concerned about possible detrimental effects because, at the time, disposal of the Spar *in situ* was being considered. The water around the Spar was relatively shallow in comparison with the North Feni site, and hence gave cause for concern. 'This was a preliminary opinion based on information at the time', claimed the spokesperson. 'The clear decision reached in the end was that deep-sea disposal would be in accordance with international conventions'.

SHELL'S CHANGE OF HEART

On Tuesday 20 June at 6 p.m., in an eleventh-hour decision, Shell eventually gave in to the pressure to reverse its decision to dump the Spar. A statement from the company read: 'Shell UK has decided to abandon deep-water disposal and seek from the UK authorities a licence for onshore disposal'. It went on to say that:

> This application for onshore disposal will include a further review of methods to minimise the risks involved....Shell UK Ltd still believes that deep-water disposal of the Brent Spar is the best practicable environmental option, which was supported

by independent studies. Shell obtained the necessary permit for disposal from the United Kingdom authorities which have handled every aspect of the approval process in accordance with established national and international policies and standards.

Less than two hours earlier, during Prime Minister's Question Time in the House of Commons, John Major had been re-affirming yet again his support for Shell's decision to dump.

It was later reported that on that Tuesday morning, the committee of Royal Dutch/Shell's managing directors met in their headquarters at The Hague for their regular weekly meeting. Chris Fay, Shell UK's Chief Executive, was also in attendance. Discussion was dominated, however, by the ongoing implications of the continuing dispute. It was thought that the heads of Shell's subsidiary operations on the Continent expressed concern over the sustained boycott of Shell products and more worryingly, in Germany, the attacks on petrol stations and letter bombs to station managers.

In a hastily arranged press conference at 7 p.m. on Tuesday evening, Chris Fay told the awaiting journalists that Shell had decided to seek to dismantle the installation on land. He said it had been ordered to climb down by its parent company, the Royal Dutch/Shell Group. The group's top management had been forced to abandon deep-sea disposal because European subsidiaries of Shell were 'finding themselves in an untenable position'. He admitted that it was an embarrassment for the company but denied that Greenpeace and its campaign had had anything to do with the reversal of its decision, 'I can say quite unequivocally that our decision has nothing to do with the four Greenpeace people sitting on the Brent Spar'. He went on to blame the European governments that had pressured both the company and the British Government, but was careful not to put the blame on any one in particular. 'Shell group has had to react to its failure to persuade ministers in certain European governments to adhere to treaties they are party to. If that is an embarrassment, yes I am embarrassed', he said. He went on to say: 'this is the first example where governments have openly protested against an option which has been carried out in a lawful and proper manner'.

Fay went on to talk about the difficulties that lay ahead for the company in trying to change its plans at this late stage. He did not think that it would be straightforward to get the British Government to agree to onshore disposal: 'I don't believe a licence from the UK Government for onshore disposal is going to be an easy process'.

The first thing that had to be done was to find a suitable resting place for the Spar until its new fate could be decided. At least 180 metres of water were needed for anchorage, and at the time of the press conference Fay was unsure of where that was going to be: 'I am not sitting here tonight knowing where I can tow the Spar to'.

Back in the Atlantic Ocean the Greenpeace protesters first heard the news over the radio at 6 p.m. Cheers were heard and the *Solo* sounded its tug's foghorn in celebration. The four men on the Spar climbed to the top of the structure and unfurled a Greenpeace banner. The water cannon, still in use to protect the Spar from more unwelcome guests, were withdrawn and the Shell vessels moved away.

It seems the turnaround came just in time for Greenpeace. It later emerged, and was confirmed by Armed Forces Minister Nicholas Soames, that the Ministry of Defence had sent a small contingent of the Commando group of the Royal Marines to the Outer Hebrides, allegedly to help Shell seize control of the Spar. A twenty-strong team would have used a technique known as 'fast roping', where they would descend from hovering helicopters onto the platform below. The Commando group are specially trained in oil-rig protection. They keep detailed plans of all North Sea rigs just in case one is taken over by terrorists, and they regularly practise assaults of rigs and rescuing 'hostages'.

Mr Soames admitted, in a Commons written reply on 14 July, that the contingency arrangements were thought of 'in anticipation of a possible police request for service logistical assistance, and a Royal Marine detachment and supporting elements were deployed to Kinloss against this contingency'. Greenpeace commented in a statement: 'It is rather worrying that the Government was prepared to use the military against peaceful, non-violent protesters'.

The tugs pulling the Spar slowly changed direction and headed back towards the North Sea. Although it had no particular destination, the Spar had to be continually towed to avoid it toppling over. The first task for Shell was to find a haven in a deep-water anchorage. There were some suitable anchorages in Scotland, but finding a place where it could be suitably dismantled would be much harder, claimed Shell. The danger of the Spar toppling over in shallow water, and the contaminants entering the surrounding waters, was a particular concern, especially through the stormy winter months in the North Sea.

Back at Whitehall, the British Government did not take kindly to this abrupt turnaround by Shell. The Prime Minister was said to be very angry and embarrassed at the company's U-turn. Michael Heseltine, the then Trade Secretary, in an interview for *Channel 4 News*, attacked Shell for backing down. 'They should have kept their nerve and done what they believed was right', he said, and added: 'I think the Prime Minister has behaved in an exemplary way and he deserved better from Shell'. Tim Eggar said on Sky television that Shell would have to work hard to find solutions to the problems that had highlighted the reasons why the Spar should have been dumped in the ocean in the first place. He went on to say, in stronger terms: 'I'm not going to imperil the UK environment just because Shell have changed their mind. They have given in to what can only be described as blackmail'.

Greenpeace then began to wave a consolatory white flag. Ulrich Jurgens, Campaigns Director for Greenpeace International, said consolingly: 'It's not a victory against Shell or the British Government. It's a victory for the sea'. He went on to add: 'We're all willing and eager to work with industry and the Government to find the real best practicable environmental option'.

Details then began to emerge about why Shell had been so reticent to dispose of the Spar on land. It was stated that the platform had gashes in its side and two of the six storage tanks had filled with seawater. The draught of 300ft made it impossible to get the installation ashore without first turning it on its side – with an inherent risk of it falling apart with the stress. It was estimated that it would be six times more hazardous to dismantle it on land than at sea. Once it was on land, there would be a risk to dismantling personnel of radioactive dust from the scale that had formed inside pipes and tanks. The steel in the structure would then be cut up for scrap. The problem of what to do with the toxic substances contained within the structure would then have to be addressed – specialist handling and disposal would be needed and the resulting material could finally find its way into landfill sites.

A Gallup survey held around the time of the Spar conflict showed an increasing concern among British consumers for environmental issues. Up to 60 per cent of those surveyed stated that they would be prepared to boycott products and stores on ethical grounds. Indeed, the poll found 33 per cent of the 30,000 questioned included moral issues in their shopping choices. 'More and more people are now realising that they can be a force for change and consumers are beginning to flex their muscles', said Terry Thomas, the managing director of the Co-operative Bank, one of the few high street banks that put ethical issues to the front of their agenda.

By Wednesday 21 June, an embarrassed and angry British Government was warning Shell that it might not allow the Spar to be dismantled on land. On BBC television, Tim Eggar said: 'Shell have admitted they are now seeking the second-best environmental option. That is no way for a major British company to behave'. Eggar also noted that the European states that had come out against dumping so vocally in the last month had known about the plans since February, but had only felt moved to oppose this option when Greenpeace began its actions in May. He added that Shell would not get any taxpayers' money to help with the extra costs of dismantling on land even if they did persuade him that it was the next best option. As companies could offset between 55 and 70 per cent of the cost of disposal against tax, the onshore disposal cost of £46 million could have cost the British taxpayer approximately £25 million, instead of about £6 million payable for offshore disposal. 'I have written to Shell', said the Energy Minister, 'to say that, in view of the fact that they recognise that deep sea disposal is still the best environmental option, I see no reason why the tax payer should pay for the second best option'. It was made clear by officials however, that Government would actually have to legislate to block the tax refund on the additional costs. Unless, of course, Shell decided to forgo the claim voluntarily.

Shell wrote personally to John Major, who had referred publicly to the Shell bosses as 'wimps', to offer its 'most sincere apologies' for embarrassing him. John Jennings, Chairman of Shell Transport and Trading plc, wrote to Major regretting the 'position in which the outcome of this complex issue placed you and your ministerial colleagues'. This letter did not cool Major's anger at the betrayal: in Parliament he warned that European Union leaders would pay for their weakness over this issue. 'I deeply regret Shell's decision to cave in to misguided pressure from environmental groups and from foreign governments', said the Prime Minister. 'I shall make it clear to my colleagues across Europe there will almost undoubtedly be a price to pay for their weakness in this particular respect'.

The Opposition milked this embarrassment for all it was worth. Ministers were out of touch with best environmental practice and voters in Britain as well as on the Continent, they alleged. Labour's

Frank Dobson urged both Government and Shell to 'come clean' by publishing all of the documents and 'independent' reports that they claimed had influenced their original decision. 'If either side refuses, it must be because they have something to hide', he said.

The MP for the Western Isles, Calum MacDonald, said he did not think that another oil company would now dare attempt to dump another platform at sea. He said:

> For a £30 million saving in disposal costs Shell has attracted hours and weeks of the worst publicity any multi-national company has had. Even before it decided the about-turn the battle had been won. No other company is going to put its head into such a noose.

He went on to make the poignant comment: 'Recycling the metal on the Brent Spar would be the equivalent of 400 million beer cans. How can the Government ask people to recycle their beer cans for environmental reasons when they are prepared to support such environmental waste?'.

However, on 22 June, another momentous news story took over from the Brent Spar saga and drove it from the headlines for a few weeks to come – John Major resigned as leader of the Conservative party. The following Monday, John Redwood announced his candidature for the post and so began the two-horse race which was to dominate the media until Major's eventual victory on 12 July.

In the City, the Spar fiasco did little to dent Shell Transport and Trading Co. plc shares, however. By the end of trading on Monday 26 June, less than one week after Shell's U-turn, its shares had risen by 4p (to 753p) whilst all around it the top 100 shares fell in value, with the sole exception of Asda which ended unchanged. It was considered that investors, mindful of the dominance of Shell's Dutch partner, Royal Dutch, in the decision to dump the Spar, and the fact that Royal Dutch shares were at an all-time high against Shell, were buying into Shell as a cheap way into the group as a whole. Before this time Royal Dutch's shares had stood at a lower rating than those of its UK counterpart, but by the end of the week the premium of Royal Dutch shares to those of Shell was about 10 per cent, very close to a historic peak.

At the end of June it also became apparent that the tide was turning against Greenpeace with regard to their scientific argument against dumping. A controversy surfaced, regarding Greenpeace's claim that the Spar's toxic contents would damage marine life, from two geologists based at Royal Holloway College, London. In a paper published in the science journal *Nature* (Nisbet and Fowler 1995) Professor Euan Nisbet and Dr Mary Fowler put forward the hypothesis that, far from damaging marine life, the Spar's contents could possibly be beneficial to it.

The scientists stated that the concentrations of toxic substances on board the Spar would have been minuscule in comparison with the quantity of natural substances released into the sea by under-water geysers. Within the Spar it was estimated that there were hundreds of kilograms of substances such as cadmium, zinc, lead, nickel and mercury. These amounts compared with between 500,000 and 5 million tonnes a year estimated to be released from the Mid Atlantic Ridge in one year. The Guymas Basin in the Gulf of California, for example, has a seabed laden with naturally deposited heavy metals and a thriving biological community exists around many metal-rich environments. 'For some ecosystems a supply of heavy metals is essential', the researchers claimed.

The geysers, also known as hydrothermal vents, are home to unquantified numbers of bacteria, and a whole range of fascinating animals that thrive on the special conditions in these areas. Nisbet and Fowler claimed that, in a worst-case scenario, Brent Spar would do nothing to harm marine life in these areas and at best, might have acted as a man-made vent, spewing out its contents for the marine life present to feed upon. These findings were based on research carried out on hydrothermal vents found in the deep waters of the Atlantic and Pacific Oceans.

The scientists compared dumping the Spar at sea with the risks of dismantling on land. They concluded that the heavy metals involved in the clean-up operation could do more damage in a land-based environment than they could ever do in a deep-sea burial site.[3]

On 28 June, the signatories to the Oslo-Paris Convention met in Brussels to discuss sea dumping of oil installations. At the meeting the Commission voted for a moratorium on the dumping of all disused oil installations. Only Britain and Norway, the two countries with large North Sea oil installations coming to the end of their useful lives, opposed the Decision. Although the Decision did not bind the UK, there would nevertheless be repercussions. Thirteen deep-water platforms in the UK sector of the North Sea will reach the end of their useful lives by the turn of this century.

Previous OSPAR rules allowed their disposal to be decided on a case-by-case basis, with provision for consultation between member countries where it was proposed that a platform should be abandoned or dumped at sea. The new Decision guaranteed an international debate every time the UK signalled its intention to dispose of a platform in this way.

Chapter 7

The aftermath:
July–August 1995

By July, the Major–Redwood battle was in full swing and dominated column feet rather than inches. The media hysteria and international outcry regarding the Spar had now subsided. May and June had seen unprecedented media coverage, the like of which had never before been seen for a single environmental issue. For example, in the UK press alone, there were over 300 articles in the broadsheet press concerning the Spar in the two months following the original occupation, and over 1,200 between May and November 1995. This is put into context when the press coverage of other environmental issues is examined: the Newbury by-pass campaign, in 1996, resulted in 165 articles between January and March; the Sea Empress oil spill drew 255 articles in February 1996; and, following in the wake of the Spar, the protest over the French resumption of nuclear testing in the South Pacific, drew in approximately 1,000 articles over a six-month period between June and December 1995.[1]

Environmental news stories in August were focused on the French government and its desire to conduct further nuclear testing in the South Pacific. The Germans, fresh from their successful and vociferous role in the climbdown by Shell, were turning their attention to the environmental 'atrocity' about to take place on the other side of the world. A boycott of all things French loomed. Restaurants said they would no longer serve French food or wine and holiday-makers were advised not to venture to France for their annual vacation. However, the German government did not seem as keen to voice its concern publicly over nuclear testing as it did over an oil platform's disposal. Even Greenpeace seemed a little more reticent, with Thilo Bode, head of Greenpeace International, saying: 'nothing would be worse than a national protest movement in Germany against France'.

Chancellor Kohl, so outspoken in his disgust over the Spar's deep-sea fate, said that the French testing was an 'internal matter' and would not be drawn any further. Greenpeace too, were against a boycott, instead they were urging Germans to write to Chirac to voice their opposition. 'It's absurd to punish a vineyard-owner because France has a new President', said Thilo Bode.

On 7 July, Norway agreed to store the Spar in a fjord, for up to one year, while Shell decided on an alternative plan to dispose of it. The Norwegian Minister of Industry and Energy, Jens Stoltenberg, said that his government had agreed to store the Spar in a deep inlet on the west coast. 'I am glad that the Norwegian government has been able to help Shell UK find a solution for storing the Spar', he said. Conditions of the agreement meant that Shell had to gain approval from three different Norwegian authorities before it was granted permission to bring the Spar into Norwegian waters. First the Norwegian Fisheries Ministry were consulted before the Spar could be stationed offshore. Then the Defence Command of Southern Norway would need to give the all clear for the Spar to arrive at Erfjord, and finally the Norwegian State Pollution Control Authority would need to be satisfied that there was no risk of pollution before the Spar was moored. Before it could moor the Spar, Shell had to carry out maintenance work that involved removing the explosive charges placed on the structure to sink it, and dealing with the two anchor chains which were hanging freely. The Spar reached Erfjord (the inlet where, incidentally, it had started out on its North Sea life back in 1976) on 10 July and was anchored there safely without further incidents. It was to remain in its temporary home for well over three years (see figure 2).

At the same time, Shell announced that it had contracted an independent inspection agency, Det Norske Veritas Industry (DNV) to investigate and produce a report on the inventory of the Spar, at a cost of 2 million Norwegian kröner (£202,000). This was ordered partly to 'allay any concerns over alarmist claims made by Greenpeace over recent weeks', a statement said. It would also provide a valuable reference point for further analysis and comparison of disposal options. DNV would also ask Greenpeace to supply them with the evidence the group claimed to hold concerning the concentration of the toxic waste allegedly inside the Spar. Greenpeace appeared to welcome the announcement and said: 'Obviously we welcome the fact that an independent body has been invited to oversee this work. It is what should have been done a long time ago because Shell is only guessing at what is on board'. The audit commenced on 4 August and was expected to take some weeks to complete.

Back in Britain, Tim Eggar continued to insist that the Spar could still be dumped at sea. He told the Commons, on 12 July, that future methods of disposal would be studied and that new reports made on the environmental impact of the Spar's disposal would be made public. As far as he was concerned, the only victims of deep-sea disposal would be the 'worms on the sea bed' (see Chapter 3, n3). His belief was given a boost when, on Tuesday the 18th, British Government NERC scientists told a parliamentary briefing that, on balance, deep-sea burial was still the best environmental option for the Brent Spar if the environmental arguments were taken in isolation. Professor John Krebs, Chief Executive of the NERC and leading the briefing team, said it would be a mistake to close the door on sea dumping of other North Sea oil and gas structures. He went on to say that the sinking of the Spar in approximately 8,000 feet of Atlantic ocean would have caused limited environmental damage, all confined to the deep ocean. Worms, molluscs and crustaceans over an area equivalent to two football pitches would have been wiped out by the impact of the plunging Spar and the spread of sediment. The Professor went on to say that, even if Shell had underestimated by a factor of ten, the quantity of oil, radioactive salts and toxic metals inside the Spar, deep-sea dumping would not pose a significant environmental threat. The research council had arranged the briefing session to help explain to ministers, MPs and peers the basis of the scientific evidence for the Spar's dumping options. The Council had called the briefing because, despite being the employer of the bulk of the country's deep-sea scientific expertise, they were concerned that they had not been consulted by Government when the decision to dump the Spar was first discussed and agreed upon (see Chapter 3).

Tim Eggar welcomed the scientists' opinions, and at the briefing said that their findings vindicated his government's position. 'Shell's decision not to proceed with deep sea disposal was regrettable... Deep-sea disposal of the Brent Spar is the best practicable environmental option', he said. This briefing only helped to heighten his resolve that Shell would not have an easy time convincing Government that onshore disposal was a viable and safe solution. 'It should not be assumed that the option of deep sea disposal has been eliminated', he said. Shell also welcomed the report, but insisted that their position had not changed because of it. They would still look into disposal on land as the second best option.

New Scientist, on Saturday 15 July, carried an article by Fred Pearce on Brent Spar. In it he called on Greenpeace to 'come clean about their motives'. Pearce pointed out that although Greenpeace had used scientific arguments against the dumping of the Spar, as they had in previous campaigns such as those against whaling, no amount of research demonstrating that their case was unfounded would persuade them to accept deep-sea disposal since they are opposed to it on moral grounds. Although the overwhelming weight of so-called 'expert opinion' was by now reported as broadly supportive of the deep-sea disposal option, at least from the point-of-view of potential environmental impact, Pearce acknowledged the validity of the moral case, but suggested that its adherents, including Greenpeace, should openly admit this and 'forget the scientific trappings'.

In late July, it was also disclosed that Shell's top public relations executive, Mr Tony Brak, would leave the group at the end of the year. A report in a German magazine, *Manager on Friday*, that he was being fired because of the Brent Spar PR disaster was hotly denied by Shell, who said that Mr Brak was taking advantage of an opportunity made available through group reorganisation to take early retirement.

On 8 August, it was announced that the final police bill for handling the Brent Spar affair was estimated at £20,000, all of which would fall on the shoulders of the Grampian region and none on either Shell, or, more particularly, Greenpeace. Similarly, several of the national newspapers reported that Royal Dutch/Shell were re-entering Formula One car racing sponsorship by supporting Ferrari, who had just announced that they had lured the champion German driver Michael Schumacher away from Benetton-Renault. Shell denied that it had wanted Schumacher 'under its name' to help counter the loss of public support in Germany as a result of the Spar affair.

Also in early August, two NERC Rockall Trough experts, John Gage and John Gordon, had written to Shell and Greenpeace pointing out their own criticisms of the BPEO documents on which the issue of the deep-sea disposal licence had been based (see Chapter 3). They eventually published a short version of their concerns in *Scotland on Sunday* in September, and a longer version in the specialist journal *Marine Pollution Bulletin* in December

(Gage and Gordon 1995). However, long before these articles appeared, the story had broken in much more accessible sources, partly as a result of Greenpeace releasing the Gage and Gordon letter and their own follow-up letter to Shell claiming that the company had 'got its science wrong' and had 'misled the UK government'. The first public reports in the dailies stressed the fact that British scientists with the most relevant expertise had apparently not been consulted by Shell or the company's advisers. The following week's *New Scientist* carried a more detailed account by Fred Pearce of the Gage and Gordon criticisms of the BPEO documents, and an editorial intimating that there was a fundamental difference in opinion amongst the relevant scientists about the facts of the case (but again see Chapter 3 for the facts).

To Greenpeace, the Gage and Gordon intervention must have come like manna from heaven. Not only did it cast serious doubt on the validity of the scientific case for deep-sea disposal, it also came from the biologists recognised as having the best knowledge of the proposed disposal site, one of whom had originally been critical of Greenpeace. Moreover, it was from NERC scientists who had not been at the parliamentary briefing and could be construed, like the *New Scientist* editorial, as negating the NERC line given at that meeting.

At the end of August came a potentially damaging public relations setback for Greenpeace. At the International Television Festival, held in Edinburgh, delegates were told that broadcasters had been manipulated into giving Greenpeace favourable coverage during the Brent Spar siege. Emotive film of the occupation helped whip up sympathy and support for the group during its campaign and resulted in somewhat one-sided coverage from the BBC, ITN and Channel 4 news (produced by ITN). The news stations admitted that they felt exploited and embarrassed by the pressure group. The organisations stressed the need for clear ground rules regarding coverage of future Greenpeace campaigns to ensure that both sides of any debate were given proportionate exposure. They vowed to take more care over future reporting of Greenpeace activities. This news was unfortunate for Greenpeace, as it was in the middle of its campaign against the French nuclear tests at the time. The TV channels were determined that the same manipulation of the press would not occur over coverage of the Mururoa Atoll nuclear tests.

Greenpeace had supplied the news agencies with its own coverage of the Spar occupation, this being gathered from its 24-hour news department equipped with film crews, editing suites and satellite technology; and from where footage was then distributed to the various agencies with no charge for usage. News stations admitted that they had relied too heavily on the emotive reporting by Greenpeace and had not given enough coverage to independent scientific evidence which tended to favour Shell's argument for scuttling the Spar at sea. However, in all fairness to Greenpeace, both the BBC and ITN had turned down invitations to send out their own camera crews to the Greenpeace vessel *Solo*, which was shadowing the Spar on its final journey.

David Lloyd, Senior Commissioning Editor of News and Current Affairs for Channel 4, proposed giving out 'health warnings' regarding footage supplied by pressure groups. Lloyd admitted that he felt 'bounced' by Greenpeace in Channel 4's news coverage of the event. 'By the time the broadcasters tried to introduce the scientific arguments, the story had long since been spun in Greenpeace's direction', he said. 'When we attempted to pull it back, the pictures provided to us showed plucky helicopters riding into a fusillade of water cannons'.

Richard Sambrook, in charge of BBC television and radio newsgathering operations, told delegates: 'In some senses we were had over Brent Spar. It was our own, the media's fault, and it is time we woke up....I'm left feeling Greenpeace were pulling us by the nose through too much of the campaign'.

Video News Releases (VNRs), films and commentary sent out by lobbyists and pressure groups to highlight their cases, are growing in use by TV newsgathering stations and are the television equivalent of the press release. Sambrook stated that he thought the growing propensity to use this material was 'an abdication of journalism'. He went on to talk about their experience of companies who would refuse permission for news journalists to film on the basis that they have VNRs already made which they would be happy to supply to the journalists. He also suggested that there were signs of this behaviour spilling over into interviews. It was suggested that VNRs from such groups should be banned unless the integrity and uniqueness of the footage could first be established before it is shown; its source should also be made apparent by clear labelling.

Richard Titchen, Director of Public Relations for Greenpeace and an ex-BBC journalist of sixteen years' experience, said he would be happy to consider using a pooled cameraperson selected by the broadcasters themselves on future campaigns, but criticised the news companies for not taking the initiative themselves. Titchen later went on to explain how the Greenpeace communication network worked. Greenpeace Communications has a list of freelance photographers and camerapersons around the world. Any footage could be transmitted, even from the remotest locations, back to London via satellite telephone link using a 'squisher' – a device that converts footage into a stream of digital signals, which can be sent by the satellite telephone link to its headquarters at a fraction of the cost of other satellite transmissions. The footage is then converted into a VNR with accompanying script and voice-over. The entire footage is appended at the end of the VNR so that the individual news agency can perform its own editing if necessary.

Greenpeace also fought back against this media attack with an offensive of their own through an article in the *Independent* of 29 August by Chris Rose, Campaigns Director of Greenpeace UK. He spoke of Greenpeace's amazement and anger by the claims at the Edinburgh Festival that the news producers felt 'conned' by the Brent Spar campaign. He claimed that ITN and the BBC were invited onto the Spar in the early stages of the campaign, but both companies declined the offer. He said that other journalists and broadcasters, who did send reporters, did not complain about their dealings with Greenpeace or about any limitations on their freedom to cover the event. He argued that if a news programme declines to cover the event personally, it was totally unfair to criticise the group for offering to supply film footage of their own.

Rose went on to report that media organisations had been offered a place on the Greenpeace vessel in the South Pacific. As there were strict restrictions on size and space, only one person from each media organisation was allowed access. However, that person could either be a reporter or a cameraperson and if they wanted to transmit their own pictures back to the UK they would be free to do so. He also defended the claims that Greenpeace had obscured the scientific arguments surrounding the Spar:

> Brent Spar aroused strong emotions not because of poor science but because the UK Government and Shell Oil defended their own interests using science as a false justification. Greenpeace, along with most international governments and millions of people, believe it is wrong to dump industrial waste at sea and that point of view deserves as much coverage as any other.

And with a very neat and justifiable soundbite, he finished the article: 'The media got drunk on the drama of the Brent Spar story and now they blame us for the hangover'.

In a BBC film shown on Sunday 3 September, *The Battle for Brent Spar*, it was made clear that Shell had not given up hope of dumping the Spar at sea. Chris Fay, Shell UK's Chairman said in the film: 'You can't rule out any option. For example, in America there was a ban on all dumping and you now have a situation where jackets of rigs are being dumped not in the deep Atlantic but in coastal waters'. Tim Eggar also added his opinion that deep-sea dumping was still the best option: 'Certainly, as of this moment I have no doubt about that. That is what we have spent three and a half years establishing'.

Greenpeace's next unfortunate piece of publicity came from an unexpected source: itself. At the beginning of September, Peter Melchett sent a letter to Chris Fay, admitting that they had got their estimates of residual oil levels in the Brent Spar storage buoy wrong. It now appeared that the Greenpeace claim of 5,500 tonnes of residual oil was vastly overestimated. Greenpeace admitted that, during its occupation, the probes it lowered into the tank to test the levels did not actually reach inside but stuck in the inlet pipe where oil did remain, giving a false reading (see Chapter 1). Melchett wrote in his letter:

> We have realised in the past few days that when the samples were taken the sampling device was still in the pipe leading to the storage tanks, rather than in the tanks themselves....In many references to our sampling, we stressed that the results were not definitive. But I'm afraid that in writing to you and your colleagues on the Shell UK board on June 19, I said our sampling showed a particular quantity of oil on the Brent Spar. That was wrong, and I apologise to you and your colleagues for this.

Melchett did, however, go on to say that he still thought Shell's scientific evidence justifying the dumping of the Spar in deep ocean did contain fundamental flaws.

Greenpeace was, understandably, trying to play the whole issue down. It had been, it said, simply an error of interpretation, and it did not alter the fact that there were lots of other uncertainties about the contents of the Spar that would only be resolved by the

DNV survey, and would never have come to light if the structure had been sunk as Shell intended. In any case, it made no difference to Greenpeace's argument against the deep-sea dumping since its objections had been based on the principle of using the sea as a waste dump and not on the contents of the Spar. Moreover, compared with the gross errors made by Shell and the Government, including not even bothering to consult the scientists who knew most about it, Greenpeace's mistake was trivial.

This revelation came as the Offshore Europe '95 conference was ongoing in Aberdeen. Chris Fay, speaking at the conference, criticised activists who staged campaigns 'based on publicity stunts and wild unsubstantiated allegations', although he did not refer specifically to Greenpeace itself. He went on to say, at the opening of the conference, that 'we know how seriously we take our responsibilites towards the community and the environment. But perhaps we have too easily taken for granted that the public understands these things'.

At the conference Dr Fay presented a paper entitled 'A vital interest: the contribution of the offshore oil and gas industry'. However, subsequent questioning by participants after his speech concentrated almost entirely on the Brent Spar issue and its likely outcome. The line of questioning was in no small way influenced by Greenpeace's recent apology.

Dr Fay reported that the Brent Spar debate had shown that the public needed to have a better understanding of the oil industry, its standards and values. He stated that 'The offshore oil and gas industry – in Europe and elsewhere – had always known it had a duty to deal properly with the installations when they were no longer required'.

Dr Fay welcomed the apology from Greenpeace which, he said, confirmed the care Shell had taken over cleaning the platform in the first place, and said he respected Greenpeace for admitting it was wrong. He also announced that Shell was now beginning a detailed review of the new proposals for disposing of the Spar and that, in light of Greenpeace's mistake, he was about to have face-to-face discussions with Lord Melchett on the abandonment of the platform. Shell was considering approximately 200 options for dismantling the Spar, and deep-sea disposal would remain one of these.

Shell were clearly moving into an image rebuilding mode, a mode they were at pains to keep up through the following three years. They had already issued a glossy brochure in late July more or less admitting that they had made a mess of their PR with the Spar and intimating that they would do better in the future. They knew that within a couple of weeks they would be launching their 'The way forward' campaign (see Chapter 8). This was no time to rock the boat with Greenpeace, especially since Fay and Melchett were due to have a face-to-face discussion on the issue of the Spar's future.

Greenpeace's apology was seized upon by Government, however, who had no such qualms, and who accused Greenpeace of scare-mongering. Tim Eggar said that Greenpeace's mistake had swayed European governments against the British decision. 'Greenpeace were making wild allegations and they were not prepared or willing to debate the scientific facts', he said. Lord Melchett responded to this with the argument that the measurements did not play a signifi-cant part in the campaign, as they were not available until the siege had run its course. And, anyway, the campaign had always centred on the possible precedent that the dumping of the Spar at sea would cause. Greenpeace also denied that the apology had dented its image: 'Obviously it is regrettable we made the mistake. But the reason people support us is even when we make a fairly small mistake like this we are prepared to admit it', said a Greenpeace spokesperson.

The independent audit of the contents of the Brent Spar, carried out by DNV, was presented to the UK authorities, Shell and the public in October 1995 (see Chapter 1). As we have already seen, DNV's findings broadly confirmed Shell's original estimates. The report concluded by saying

> On the basis of the investigation carried out by DNV it has been demonstrated that Shell have underestimated the quantity of hydrocarbons remaining on the Brent Spar after decommis-sioning. The origin of this underestimate is almost certainly because Shell failed to appreciate the persistence of hydrocar-bons clinging to the internal surfaces of the piping system and tanks. The vast majority of the total hydrocarbons identified can be removed by relatively simple methods.
>
> From analysis of sludge and water samples taken from the storage tanks DNV conclude that the quantities of heavy metals remaining on the Brent Spar are in broad agreement with those estimated by Shell. These quantities of heavy metals are unlikely to pose an environmental hazard dependent on the final disposal methodology adopted. By measurement and calculation of the

sum total radioactivity of all naturally occurring substances aboard the Brent Spar it is shown that the original Shell estimates were correct. The sum total radioactivity aboard the Brent Spar is similar to that to be found naturally occurring at many onshore sites.

From the inventory investigation an additional quantity of potentially damaging PCBs have been tentatively identified. The existence of the components containing these PCBs can be verified easily and removal effected simultaneously for professional disposal.

This report must have come as a great relief to the Shell personnel involved with the decommissioning project. With their residual oil calculations verified and an official apology from Greenpeace they must have thought the tide was turning in their favour. But they still had a long and difficult road ahead.

Chapter 8

The way forward
The search for a solution

Following the decision to abandon the deep-sea dumping operation, Shell received many unsolicited suggestions for how they might dismantle and dispose of the Spar, nearly 200 by the end of July and ultimately almost 450. They ranged from the totally conventional, through the imaginative to the outlandish, including turning it into a hotel or casino, using it to generate wind or wave power, sinking all or part of it to form an artificial reef or fish farm, and incorporating various bits into harbour, breakwater or bridge building projects.

The strength of the response, along with the events which had led to the capitulation in the first place, confirmed what Shell must have known since early May: that it had made a total mess of the public relations surrounding the Brent Spar. In the words of Heinz Rothermund, Managing Director of Shell Expro, 'We have acknowledged that we originally set out to dispose of the Spar without explaining what we were doing early enough or widely enough'. Accordingly, as we have already seen, Shell embarked upon a major campaign to attempt to correct the situation. The objective, again to use Shell's own words, was unashamedly 'winning hearts and minds', but the broad intention of its 'The way forward' campaign was not made public until October.

THE TRUTH ABOUT THE CONTENTS

There were several elements to the new strategy. First, the controversy which had raged over the nature and contents of the Spar had to be settled. Shell had already commissioned Det Norske Veritas to carry out an independent survey as the Spar lay in Erfjord. As we saw in Chapters 1 and 7, when the resulting DNV Report was made public later in October 1995 it largely exonerated the claims that Shell had made and contradicted those of Greenpeace. Shell must have been fairly confident of this outcome before it placed the contract, so by the time it announced its 'The way forward' campaign it was reasonably certain that this aspect, at least, would not cause it any significant embarrassment.

THE STRUCTURAL INTEGRITY

A second element was to re-assess the Spar's integrity in relation to the engineering problems which might now have to be faced, such as rotating it to a horizontal position or raising it out of the water vertically. So Shell commissioned a specialist engineering company, W. S. Atkins, who had already contributed background documentation for the deep-sea disposal BPEO, to conduct the new survey. The resulting analysis, claimed Shell in a press release of 3 July 1996,

> shows that reversing the original installation procedure to bring the Spar ashore for dismantling is far from straightforward; [and] that raising the Spar from the sea presents a huge challenge in overcoming major structural limitations and safety risks....The Spar's storage tank walls are thin, stiffened membranes, designed to ensure they do not experience high external water pressures while floating in the normal position – an efficient and safe design for the hostile waters of the northern North Sea. But the new analysis shows that the Spar was highly stressed during installation, close to the point of failure. Safety regulations mean that it cannot be placed under these same stresses again, which could make the tank walls buckle or collapse.

Clearly they were saying that the goalposts had moved even since they began the decommissioning process.

WHO WILL RID US OF THIS TROUBLESOME BEAST?

A third element was to find a suitable contractor capable of carrying out whatever the ultimate solution turned out to be. So in

October 1995 Shell placed a Periodic Indicative Notice or PIN in the *Official Journal of the European Communities* inviting qualified potential contractors to submit their credentials so that Shell could select those who would be invited to proceed to the next stage. This process was fairly drawn out and the names of the twenty-one companies that passed this first screening, what became known as the 'long list', were not announced until July 1996. Each contractor was provided with all the previous studies on the Spar, including access to the ideas and suggestions received by Shell, and given until 31 July to submit their own outline concepts of what they thought should be done with the Spar and briefly how they proposed to achieve this. These outlines were to be published by Shell in August 1996 and then narrowed down to a shortlist of half-a-dozen or so which the originators would then be asked to develop into detailed concepts.

THE 'LONG LIST'

By the July deadline, nineteen of the long list contractors had responded with outline proposals, some with more than one idea. They ranged from more or less total refurbishment and re-use of the Spar, either intact or cut into pieces for different functions, to complete dismantling and disposal and recycling (see Appendix 2).

Most were for dismantling and recycling onshore, in dry docks or at least in conventional dockyards. They differed mainly in the proposed techniques for getting the Spar from its vertical orienta-tion in Erfjord to the dismantling locality. Most first involved removal of the topsides, the wide part at the top housing the accom-modation and supporting the heli-deck. This section, weighing about 1,250 tonnes, could be lifted clear using relatively conven-tional existing marine salvage and shipbuilding techniques. Having removed it, there were several different suggestions for what to do with it. Some involved simply carrying it ashore for recycling, but several potential contractors suggested that it could be refurbished in a specialised role either ashore or at sea, for example as a training centre, perhaps for the oil and gas industry. One contractor, AMEC from Aberdeen, suggested its use, perhaps standing on part of the supporting column, in a tourist facility connected to a causeway in the Morecambe Bay area, presumably cashing in on the Spar's celebrity, or notoriety.

After removing the topsides the potential 'recycling' contractors all had to face the problem of how to get the main body of the Spar, particularly the enormous storage tanks, out of the water and cut up. Some had access to major inshore shipyard-type facilities to which the Spar could be towed in its vertical orientation, despite its almost 100m draft even after removal of the topsides. For the rest, the major inshore facilities were separated from Erfjord by waters which are far too shallow to allow the structure to be towed vertically without going aground, with potentially disastrous consequences. So their solutions had either to envisage up-ending the Spar to a horizontal orientation before towing, or dismantling it in deep water by raising it vertically and cutting slices from the top as it emerged. Let us consider raising the Spar vertically first.

On the face of it this seems much simpler than up-ending it since, maintained upright, the pressure at any particular point along the Spar's length would at least be the same all the way round. But there are complications. First, the Spar weighs in excess of 60,000 tonnes with its tanks full. This is a big lift by any standards. So the obvious answer would seem to be to empty the tanks. But the tanks would have to be emptied for a quite different reason. The thinness of the tank walls means that they would be unable to withstand the pressure from more than a few metres of water. Lifting the Spar full would be a bit like filling a paper bag with water in a bowl and then trying to lift it clear; it would simply burst. Similarly, if the tanks were pumped empty while the bulk of the Spar was still submerged the tank walls would collapse inwards under the weight of the surrounding water. So the tanks would have to be emptied gradually as the Spar is raised, keeping the water levels inside and outside more or less the same at all times. Since the water in the tanks is contaminated with oily residues and sludge it could not simply be dumped, but would have to be pumped to some storage system for subsequent treatment; not in itself a difficult problem, but a further complication in an already complex operation. And the damaged tanks would have to be treated very carefully to ensure that they did not leak during lifting and cause excessive differential pressures on the walls shared with other tanks. Considering all these factors, along with the possibility of bad weather and the problem of exactly what the lifting wires or jacks would have to be attached to, and you soon realise that this is not a simple task.

Tipping the Spar to the horizontal position has its own, and in some ways even more difficult problems. First, you would need to apply a turning force, pulling the top in one direction and the bottom in the opposite direction, for example. And all the time ensuring that the Spar did not flip to one side. As the structure rotated, the six tanks would emerge from the water at different rates, the upper ones first and fastest, the lower one later and slower, and the lowest of all initially not at all. So they would also have to be drained at different rates. Add the complication of the damaged tanks and you have a non-trivial problem once more. The long-list contractors clearly had a lot to think about.

PUBLIC CONSULTATION

In the meantime a fourth element of 'The way forward' was put in hand, this one to involve a much broader based community. The objective of this element, called 'The dialogue process', was ostensibly to solicit the advice and opinion of as wide a range of interested individuals and organisations as possible, including, for example, scientists, engineers, lawyers and industrialists, as well as environmental, consumer and even perhaps religious groups, and not necessarily only from the UK but also from other European countries. Shell and Greenpeace had taken part in direct discussions as early as September 1995, followed up by a meeting in January 1996 between Dr Chris Fay, Shell UK's Chairman, and Thilo Bode and Peter Melchett, Executive Directors respectively of Greenpeace International and Greenpeace UK. But Shell wished to broaden the process.

The exercise was put in the hands of an independent charity, the Environment Council, which had 'considerable experience of helping different interest groups work together to find common ground'. In consultation with Shell, the Council decided to arrange a series of one-day seminars, the first on 1 November 1996 in London.

Accordingly, a rather strange group of some seventy invited 'delegates' arrived at the Queen Elizabeth II Conference Centre just after nine o'clock for what turned out to be an equally strange, but fascinating day. They had been chosen, presumably by Shell Expro in consultation with the Environment Council, on the basis of a combination of interest and expertise. They ranged from university academics in a variety of subjects and from continental Europe as well as from the UK, to representatives from large and small private companies, from environmental groups (including Greenpeace) and from groups representing other specific interests from trade to religion. As they drank their pre-meeting coffee they were invited to write their aspirations for the day, good or bad, on 'post-it' notes and stick these onto large boards for general perusal. This set the scene for the day, because this technique of 'posting' anonymous comments, suggestions and opinions about all aspects of the decommissioning of the Spar, was used at each stage of the day's proceedings. The resulting hundreds of little bits of sticky paper were then collected together by the staff of the Environment Council and a few weeks later emerged as a report to Shell on the day's activities. This was inevitably a difficult task, at least if the hope was to draw clear and unequivocal conclusions, since there were probably almost as many opinions as there were delegates. Only two clear messages came across. First that the delegates were almost unanimously opposed to one of the long-list ideas, a proposal to bury the Spar in relatively shallow water, an option that was shortly to be dropped completely. Second, that the dialogue process should be broadened to include many more groups, particularly in continental Europe, though the role of politicians in the process was not universally agreed.

Following the London seminar, slightly modified versions involving rather fewer delegates were held in Copenhagen in March and in Rotterdam in May 1997. These seminars considered only a list of eleven proposals which Shell had selected from the original thirty following the first seminar. As before, no clear-cut conclusions were drawn from the meetings, except that all delegates believed that the Brent Spar affair would have far-reaching consequences for industry in general, and that any final solution would have to consider public opinion as a vitally important factor. Hardly 'Dog bites man' stuff. As we shall see, a second round of seminars was to be arranged later in the year as Shell approached the end of the process.

THE 'SHORTLIST'

On 13 January 1997, that is two months before the Copenhagen seminar, Shell Expro announced a shortlist of six contractors who were still in the running. These so-called 'first round winners' between them were asked to develop in detail eleven different ideas for the Spar's re-use or scrapping onshore, and to submit commercial bids for carrying out their schemes by the end of April. Five of them were also asked to develop in detail their ideas for the challenging operation of raising the Spar from the water. Each contractor was given about £250,000 by Shell to pay for the work involved in these various activities.

According to Shell, the shortlisted proposals had been selected by applying three of the four stipulated BPEO criteria, technical feasibility and risk of failure, safety of the personnel engaged in the disposal process, and potential environmental impact. The fourth BPEO criterion, cost, would be considered at the next stage once the schemes had been properly costed. A key aspect of the selection criteria, again according to Shell, was that the shortlisted proposals should cover a wide range of potential solutions. Shell further claimed that, because the dialogue seminar delegates had felt that the company was giving more weight to technical feasibility than to some of the other issues, they had carefully balanced the detailed questions to be asked in each criterion category. They also implied that their choice had been influenced by the delegates' opposition to the inshore burial proposal and several of the re-use ideas, particularly the feasibility of the power generation proposal.

The surviving disposal ideas covered a relatively limited range. They were either for scrapping the Spar onshore in its entirety, for partial scrapping but with the topsides section re-used as some kind of shore-based training centre, and finally for the hull section or parts of it also re-used, in this case as quay extensions, as coastal erosion defence structures or in a fish farm, but all in very shallow water.

The deadline was extended to 2 June, by which time all six of the contractors had responded with detailed proposals for nine of the original eleven ideas that had survived from the previous stage. The two that had disappeared were the Kvaerner Stolt proposal to convert the Spar's hull into a dry dock gate and a Wood-GMC proposal to convert the topsides into a training centre. Details of the following surviving proposals were issued by Shell on a CD-ROM on 17 June.

Brown and Root Energy Services

This proposal involved towing the Spar in the horizontal orientation to the Barmac yard at Nigg in northern Scotland, where it would be dismantled and scrapped in a controlled environment.

First, the heli-deck would be removed and taken ashore. An A-frame would then be installed on the Spar's turntable and used in the remaining preparations for up-ending. The up-ending would be achieved by partially raising the structure vertically by displacing some of the water in the tanks with compressed nitrogen. The damaged tanks would be repaired by divers, and drain hoses and submersible pumps would be connected to the lower ends of each of the main tanks to carry the displaced water to a tanker for transport ashore. The nitrogen would be supplied to the tops of the tanks from a support vessel via hoses and an existing manifold on the Spar. The turntable and A-frame would now be removed.

Once the Spar had been raised a little over 40 metres it would be up-ended using two cables, one from the bottom of the structure to the up-ending barge and the other from the top of the Spar to a towing point on shore. The up-ending process would be undertaken in a slow and controlled manner over several days. As the Spar tipped and the upper tanks gradually emerged above the sea surface, the water would have to be drained from them unevenly to prevent the build-up of potentially damaging differential pressures. Once the Spar was almost horizontal, and the draught had been reduced to about 15m, it would be towed the 306 nautical miles to Nigg, taking an estimated three and a half days.

At Nigg, the Spar would be floated into a graving dock via a caisson gate with a maximum of 13 metres of clearance at high tide. So the Spar's draught would have to be reduced to less than this figure at the end of the tow. It would be manoeuvred through the dock entrance using additional tugs and a winch system within the dock where it would be positioned over a prepared support system. The dock gate would now be closed and the dock drained. Any particularly hazardous materials present would now be removed, if necessary within 'cocoons' with negative internal pressures. The Spar would then be cut into 1,000-tonne sections which would be moved to a purpose-built decommissioning facility where the final dismantling would be undertaken.

Kvaerner Stolt Seaway Alliance

The Kvaerner Stolt proposal involved towing the Spar from Erfjord to their yard at Hanoytangen in its vertical orientation, though it would have to be deballasted temporarily to a draught of 83m to clear the entrance. It would now be carefully deballasted and, as it rose from the water, it would be cut into five large sections. The topsides and upper part of the neck would be towed to Trondheim for refurbishment as a training facility. The main body, cut into two sections, would be transported by barge to the Karsto gas terminal where the sections would be used as mooring 'dolphins', i.e. floating structures, joined by jetties, to form part of a tanker-mooring quay extension. Alternatively, 6-metre high sections might be used in the construction of a fish farm near the warm water outfall from a Statoil refinery at Mongstad. Finally, the remaining section containing 6,500 tonnes of ballast would be towed to a dry dock for disposal.

McAlpine Doris Able Joint Venture

This proposal involved towing the structure to the Teeside Environmental Recycling and Reclamation Centre (TERRC) where it would be dismantled and major sections used in the construction of a new quay wall.

After the removal of the heli-deck and turntable the Spar would have to be reverse-up-ended for the tow. This would be achieved by displacing the ballast water in the tanks with compressed air supplied from barges moored alongside. By extracting water unequally from the tanks the Spar would be up-ended in a controlled manner until it was floating horizontally with a draught of 10m.

The three hundred nautical mile tow to the TERRC facility would take about four days. On its arrival it would be taken into a twenty-acre flooded basin which would then be drained. All mobile wastes would be removed and sent to licensed disposal facilities. The re-useable plant would also be removed and the topsides cut into sections for transport to a nearby steel recycling plant. The ballast end would also be dismantled for recycling, while the ballast itself would be removed and crushed for re-use.

The remaining column and storage tanks would be cut into rings, each weighing up to 1,000 tonnes, and transported to the site of the new quay. Here they would be cut in half vertically and the resulting semicircular structures manoeuvred into a row alongside the dock to form the basis of the new quay.

Thyssen Aker Maritime

This company's proposal involved towing the Spar to an inshore site at Hinna near Stavanger where the entire structure would be scrapped in a controlled environment. Prior to the fairly short tow the draught would have to be reduced to about 88 metres by displacing the water in the tanks with compressed air.

On arrival at Hinna the heli-deck and topsides would be removed by crane barge and taken ashore. The remains of the Spar would now be positioned within a lifting frame suspended beneath a catamaran barge and lifted vertically in steps of about 80 centimetres using twenty jacks mounted on the barge. With each lift the level of the ballast water would be reduced to maintain equal internal and external pressures. As sufficient of the Spar emerged it would be cut into sections and removed to shore for decontamination, the resulting wastes being sent to approved treatment plants. Once the base section was exposed, the catamaran would be towed ashore and the ballast removed. Finally, the re-usable components would be cut into small sections and recycled at specialist plants.

Wood GMC

Wood GMC (the ultimate 'winner') proposed the vertical dismantling of the Spar, the topsides to be dismantled and recycled and the hull sections to be used for a quay extension at Mekjarvik in Norway (figures 10–13).

Originally, the dismantling and cutting operations were to take place in Erfjord where the Spar had been berthed since the abandonment of the deep-sea disposal operation. But in early 1998 Shell decided that the structure would be moved first to a rather more sheltered site at Yrkefjord where all the necessary support vessels, lifting catamaran, tanker, transportation barges and accommodation barge would be assembled. The Spar would be raised on a lifting frame within a catamaran, as in the Thyssen proposal, deballasting to maintain equal internal and external pressures being also achieved in much the same way. As each section was detached it would be slid horizontally along skid beams on the catamaran to be loaded onto transport barges for carriage to Mekjarvik.

The sections to be removed in this way would be the topsides, weighing 1,413 tonnes, the transition column weighing 1,218 tonnes, a first main section weighing 1,814 tonnes, two further sections each weighing 1,316 tonnes, and finally the bottom section weighing 1,574 tonnes.

The transition column and hull sections would be placed in position in Mekjarvik by the catamaran and filled with sand and gravel to form the foundation of a 170-metre extension to an existing deep-water quay.

Amec

The Amec option involved the use of cut and cleaned sections being used to build reefs in a coastal protection scheme in Norfolk. It did not include proposals for raising the Spar or cutting it into sections, but only for their use once available as a result of the activities of one of the other contractors.

A long-term sea defence strategy, initiated by the Environment Agency in 1989, involves the installation of a series of inshore reefs along the coast of Norfolk to reduce coastal erosion. The current reef construction technique involves the emplacement of a core of rocks weighing 2–4 tonnes each, with an outer protective layer of larger rocks each weighing between 8 and 16 tonnes. The Amec proposal was to use Spar sections filled with sand as the core for future reefs to replace much of the rock used in the nine reefs constructed so far. The sections, each almost 5 metres deep, would be transported to the Norfolk coast by barge, placed in position at high tide and finally filled with sand and surrounded by rock armour.

THE DNV REPORT

With these broad features of the shortlisted proposals now in the public domain, the detailed documentation went through an independent vetting process before Shell's dialogue process could go into its next phase. This vetting was undertaken by Det Norske Veritas, this time employed by Shell Expro specifically to check the technical, safety, environmental and costing aspects of the proposals to ensure that all the relevant aspects had been covered and that the conclusions were sound. To achieve this, DNV had produced guidance notes for the contractors in the preparation of their proposals resulting, hopefully, in strictly comparable information. After this checking stage, DNV were to rank the various options, using published criteria, and present the results to Shell Expro. The DNV role was therefore to undertake steps three and four of the BPEO process (see Chapter 2), while the next stage, selecting the preferred option to go forward as the final BPEO for licensing, was the responsibility of Shell. This was clearly an attempt, and a laudable one, to make the comparison between the different options transparent and 'neutral', that is not directly involving Shell, with its potential vested interests, in the process.

The resulting DNV Report (Det Norske Veritas 1997) appeared in early October and was the subject of the next dialogue process seminar held in London on 15 October 1997, just about two and a half years after the Greenpeace occupation. Doesn't time fly? Over the next few weeks, similar seminars were held in Copenhagen, Rotterdam and Hamburg. The format was much the same as for the earlier seminars, consisting of a very professional presentation by Shell, this time of the DNV results, followed by group discussions of the many complex issues involved. And the result was much the same too: a plethora of disconnected thoughts, suggestions and opinions with no clear outcome. This was more or less inevitable, first because there was no obvious 'winner' from the DNV assessment, and second because there is no agreed way of comparing, for example, economic cost against environmental costs, or technical feasibility against safety of the workforce.

The DNV assessment was never intended to produce an overall ranking of the options – and it did not attempt to. Instead, as promised, it ranked them separately on technical feasibility, safety to the workforce and environmental impact, while simply verifying that the costs associated with each were strictly comparable. No single option came top of the rankings on all, or even most, of the various criteria. For example, the 'benchmark' deep-sea disposal won hands-down on technical feasibility. After all, Shell or its contractors had all but done it once already. All that they hadn't done was to fire the charges to blow holes in the buoyancy tanks and let go of the tethering lines at the appropriate time. In contrast, even the simplest of the other options, while using well tried offshore techniques, had never been used on a structure like the Spar, and particularly in its damaged and therefore fragile condition. For much the same reasons, DNV considered the deep-sea option to be the safest for the workforce, and in some cases very much safer. But even the use of this criterion is far from simple. For example, the safety hazards associated with the Amec proposal to use cut sections of the Spar in coastal erosion defences appeared to be much higher than for any of the other proposals. However, construction of the Norfolk coast barrier is a dangerous operation and will go ahead with or without bits of the Brent Spar. If Spar rings were to be used, the amount of dangerous rock handling would be correspondingly reduced and this disposal option might actually result in a net reduction in injury or death. Although cost must not be an overriding consideration in the BPEO process, the deep-sea disposal was also, incidentally, by far the cheapest at £4.7 million compared with between £11.4 million (one of the Kvaerner Stolt options) and £48 million (Brown and Root). (Interestingly, Shell's estimate of the cost of deep-sea disposal in June 1995 was almost £12 million. Has it spent more than £7 million on this option already, or has the operation turned out to be much cheaper than Shell thought?)

On the other hand, with the exception of a tiny amount of oil in the Spar which might still be recovered, if it was sunk in the deep sea all of its structural materials would have been simply thrown away. In contrast, all of the other options involved re-using or recycling much of the Spar's constituents. In assessing the net effects, DNV therefore set the 'savings', in total energy use and emissions to the atmosphere by not having to produce these materials from scratch, against the equivalent direct 'costs' of undertaking the different options. The result was that while deep-sea disposal shows a net energy consumption of some 52,000 gigajoules of energy and the emission of 4,000 tonnes of CO_2, 70 tonnes of NO_x and 3 tonnes of SO_2, the other options show net 'profits' in some or all categories. But to put these savings into some sort of context, DNV pointed out that the biggest energy savings over the deep-sea option (by the Kvaerner Stolt options) would amount to the annual consumption of only about 1,300 people in the UK, while the worst- and best case scenarios for emission costs and savings were almost all equivalent to those resulting from the annual use of no more than three or four thousand cars. The main exception was the 'saving' of some 39 tonnes of SO_2 emissions, and a net saving of 150,000 gigajoules by the Kvaerner Stolt proposals which would be equivalent to no less than 64,000 'annual cars'. Unfortunately, these brownie points were countered by DNV's assessment that they also involved intolerable technical hazards. You clearly can't win 'em all!

Without going through all the tedious detail of the DNV report in this way, the net result, which appeared to be the general consensus at the London seminar, was that there was little to choose between any of the onshore disposal options. But the delegates were not asked to make such a choice. While they were clearly at liberty to express any preference they wished, Shell were apparently more interested to know which of the various factors were considered as most important in coming to a decision, and which were not. Whether this expectation from the drawn-out, complex, expensive and time-consuming dialogue exercise was realistic or not, Shell was subsequently able to emphasise the trouble it had gone to, to identify and then to apply the value judgements of all manner of interest groups thought to be representative of the public at large. And since Shell wisely did not resurrect the deep-sea option it received no significant criticism from Greenpeace, other than the suggestion that it had made rather a meal of getting to the solution that the environmental group had advocated in the first place.

THE FINAL CHOICE

On 29 January 1998, thirty-three months after the Greenpeace occupation, Shell issued a press release and media pack announcing that they had chosen the Wood GMC proposal to use the Spar as the basis for a quay extension in Norway. It was a shrewd move, and the accompanying text followed a finely drawn path between heart-warming humility and a fairly bullish face-saving exercise.

The release was headed 'Shell chooses new life for Brent Spar: Wood-GMC's re-use as a quay. Not deep sea, not scrap onshore – a unique re-use for a unique structure'. Reinforced by subsequent statements by Shell Expro spokespersons including Heinz Rother-mund and Eric Faulds, it emphasised that the chosen route was a one-off solution to a special problem and not a precedent for other offshore structures. Thus it kept the door open, just, for other disposal options under different circumstances, including, perhaps, *in situ* toppling or abandonment and even deep-sea disposal. This would have pleased the industry and at least some sections of the DTI, particularly its offshore oil and gas office in Aberdeen.

By choosing one of the re-use options, and specifically the quay extension at Mekjarvik, Shell could also claim that this was a possi-bility which had simply not been available for consideration the first time round since plans for the extension were not then in place. The fact that this was now apparently the 'best' option did not therefore negate the BPEO process, nor was it inconsistent with the deep-sea disposal option being previously the 'best'. 'Last time', claimed the release, 'we were applying the BPEO criteria to a narrower range of options, and carrying out the detailed comparison only between an onshore scrapping option and deep sea disposal'. Though Shell did not intend this statement to be taken this way, it was certainly true that they had previously looked in detail at only a remarkably limited range of options, as we saw in Chapter 2.

The release went on to anticipate a number of obvious questions that might be asked following the new decision. The questions (with our interpretation of Shell's answers) included: 'Do other European governments have to approve your choice?' (no); 'Does this mean other offshore installations could be re-used?' (possibly, but not very likely); 'Does this mean sea disposal can be ruled out for other installations?' (society's decision, not ours, but we would be pretty hacked off if it was); 'Will you hold dialogue on all decommissioning or all company decisions?' (you must be joking!).

But in response to the question 'Have Greenpeace been proved right?' Shell's release was much more bullish and openly critical of the environmental group than it had been for many months previously. 'Greenpeace were wrong', it said, 'about a great many things'. For example:

- They were wrong about the amount of oil on board the Spar. DNV said Greenpeace made 'major errors of misinterpretation, resulting in grossly overestimated amounts of oil on board'.
- They claimed that toxic chemicals were 'hidden' on the Spar. DNV found no evidence that such toxic waste was ever hidden on board.
- They wrongly said Spar could be scrapped onshore for £10 million.
- They wrongly accused Shell of carrying out no inventory of the Spar's contents.
- They wrongly said the Shell plan was for Spar to be 'dumped in the North Sea'.
- They wrongly suggested that more than 400 oil rigs in the North Sea might also be 'dumped'.
- They wrongly accused Shell of 'extreme violence' in the peaceful operation to remove protesters from the Spar, yet saw their own campaign spark bombings and shootings in Germany.
- After being proved wrong on many facts, Greenpeace claimed that their campaign had not been about facts, but a principle. They are fully entitled to uphold a principle, but wrong to deny abuse of facts.

It is interesting that Shell should put at the head of this list the erroneous claims about the Spar's contents, irrefutable at the time they were made. It lends some credence to the suggestion that while the claims had little effect at the time on the level of public opposition to the deep-sea disposal, they may have played a significant role in Shell's senior management decision to abandon it.

Greenpeace's own press release following Shell's decision naturally welcomed it, but criticised the oil giant for taking so long to come to the obvious conclusion. It also rejected the claim that the Brent Spar solution did not create a precedent for other offshore installations. Instead, the Greenpeace release stated that

> the way is now clear for a permanent ban on the dumping of decommissioned offshore installations in the ocean. There are up to 600 such installations to be decommissioned in European Union and Norwegian waters alone, many of which are still being considered as candidates for ocean dumping by the UK and Norwegian Governments.

As far as the Brent Spar was concerned, Shell had only one more hurdle to cross, obtaining the all-important UK Government licence for the new BPEO. Despite the continued confusion about exactly what the UK's attitude to marine disposal of offshore structures now was, the change of government since the Brent Spar crisis probably made this more or less a formality. After all, the ministers involved at the time, such as Tim Eggar, who had intimated that Shell might have difficulty convincing them that deep-sea disposal was not the best option, had long since disappeared from the scene.

But before the application would be considered, the new Minister for Science, Energy and Industry, John Battle, had requested that the NERC Scientific Group on Decommissioning, the Shepherd Group, should be reconvened to vet the new option. The reconstituted Group contained most of the original members[1] but was strengthened by additional specialists, particularly engineers, to provide adequate expertise to cover the new situation. It was asked to 'comment on the adequacy of the work from a scientific and engineering perspective, carried out for Shell Expro by contractors and by Det Norske Veritas, in relation to the environmental impact of the preferred disposal option', and to 'compare the proposed option (from a scientific and engineering point of view) with the earlier deep sea option'. So once again the Group was to consider only the scientific and technical issues, and none of the other factors such as aesthetics, morals or public acceptability.

The resulting report (NERC 1998) was prepared in May 1998 but not made public until 30 June. In relation to its first task, the Group felt that some additional work should be undertaken on the details of the engineering involved in the new proposal and also on its environmental impact. Nevertheless, it was satisfied that the proposal was technically feasible and, assuming that all went according to plan, would cause minimal environmental impact. But most revealingly, the Group felt also that in the light of new information about the proposed deep-sea disposal site, particularly the relatively firm nature of the seabed, which had not been available when the first report was completed, the environmental impact of deep-sea disposal of the Spar would be even less than it had previously considered. In fact, it stated that the environmental impacts from both alternatives would be 'acceptably small, and not sufficiently different to provide useful discrimination between them' (E22, 7). In fact, purely on environmental impact grounds there is nothing to choose between the two options, so that any final choice would have to be based on other factors, including public acceptability.

Although the report is, as you would expect from such professionals, detailed and very thorough, and the Shepherd Group obviously took its brief very seriously, there is a clear, but unwritten, impression that it felt that in the overall scheme of things the question of what to do with the Brent Spar is really rather trivial. In fact, the Report implies that the whole problem of the potential environmental impact of the marine disposal of decommissioned offshore structures may have been hyped up beyond its true significance. 'On the other hand', the Report says,

> serious threats to the marine environment and its ecosystems, not related to decommissioning, certainly exist. These include habitat destruction, the impact of intense fishing activity, the introduction of non-indigenous species from other parts of the world, and in some areas eutrophication and pollution by toxic substances. These threats are generally greater than those associated with decommissioned off-shore structures.
>
> It is therefore necessary to strike some sort of balance between efforts to reduce even further those impacts which are already small, for primarily aesthetic or ethical reasons, and efforts to reduce more serious impacts due to other causes. This

is difficult, and it is not assisted if undue attention is focused on off-shore structures, and if this diverts public and political attention and final resources away from these more serious pressures on the marine environment.

(9.3.2 and 9.3.3, 30)

To help overcome this dilemma the report recommends the establishment of a UK national marine environment advisory panel 'to provide informed and open advice to the government and to the public on such issues affecting priorities for the protection of the marine environment'.

As in the case of the Shepherd Group's earlier report, this one does not express a view one way or the other about whether the Wood-GMC proposal should be adopted. But nothing in the report will prevent the government from ultimately giving Shell the necessary licence to proceed with this option. Although this book will have gone to press before the licence is issued, we are confident that this will be the final fate of the Brent Spar.[2]

But the Shepherd Group clearly felt that the Spar had revealed serious shortcomings in the way the BPEO operates, and even more significant than those identified in its first report. It probably feels that its comments and recommendations in this area will, in the long term, be much more important than those specifically directed towards the disposal of the Spar itself. We shall look at these in the final chapter when we try to assess whether the Spar affair has, on balance, been a good or a bad thing.

Chapter 9

The Brent Spar legacy

> Brent Spar has transformed our outlook. Spar is not as so many believe an environmental problem, rather it will go down in history as a symbol of the industry's inability to engage with the outside world.
>
> (Heinz Rothermund, Managing Director, Shell UK Exploration and Production, 'Matching green words with clear actions', 1997 Celebrity Lecture for the Institute of Petroleum, Strathclyde University, Glasgow, 20 May 1997)

We have no idea how many times the words Brent Spar, known to almost no-one before May 1995, have appeared in print or on radio and television programmes subsequently, but it must run into tens, if not hundreds, of thousands. By any standards, the Spar has had an amazing impact. Almost all commentators, no matter which side of the many divides they come from, seem to believe that the events surrounding it represent some sort of turning point; things will never be quite the same again. But a turning point in what? And which 'things' will never be the same again?

The first, and most obvious effects of the whole business were those on Shell Expro and its immediate parent company Shell UK. Literally overnight, instead of being able to sink the Spar, and hopefully more or less forget about it, the organisation found itself still with 14,500 tonnes of unwanted rubbish on its hands and the eyes of the world watching its every move. The immediate result was a strained relationship between the company and the UK Government, and specifically between Chris Fay, Shell Expro's Chief Executive, and the Prime Minister. But this probably blew over fairly quickly, particularly with the Major–Redwood Conservative party leadership battle taking most of the media attention from the Spar and giving Shell personnel a chance to catch their collective breath and decide on the next move. During this time they were able to find the short-term solution (at a price) of parking the Spar in Erfjord. In fact, at a press conference launching the second Shepherd Report in June 1998, Eric Faulds was quoted as estimating the total cost of finding an acceptable solution for the Spar at approximately £9 million. This sum included £250,000 each for the shortlist proposals, £500,000 for the dialogue process and the remainder spent on 'board and lodging' for the Spar and internal Shell costs. The much longer, time-consuming, expensive and sensitive task was the tortuous reassessment and public consultation exercise described in the last chapter. But this was just the tangible tip of a much larger philosophical iceberg, the chill from which penetrated all levels of the organisation. Over the succeeding weeks and months Shell produced a series of press releases and public presentations referring in one way or another to the Spar. Many of these were rather specific and came from personnel directly involved in the Spar's decommissioning. But more general statements and speeches were made by senior management, including the very top, Royal Dutch/Shell Group Chairman Cor Herkströter. They were full of hair-shirt phrases about the need to explain management decisions to the public more fully than in the past, to involve 'stakeholders' in the decision-making process and above all to 'win hearts and minds'. Clearly, Shell's image, including its self-image, had been shaken to the core. Some cynics (see Fred Pearce, *New Scientist*, 15 November 1997) even saw Shell's announcement that it was to invest $500 million in photovoltaics (the generation of electricity from sunlight) as at least an indirect response to the bad publicity surrounding the Brent Spar. Shell's attempts to distance itself from its Brent Spar partner by withdrawing from the Exxon-led Global Climate Coalition, the oil industry pressure group encouraging President Clinton not to accept the CO_2 emission limitations of the Kyoto conference, could be interpreted in the same light.

The Brent Spar was not, of course, the only public relations issue taxing Shell in 1995 and 1996. The continuing criticism of the company's record in Nigeria continued to rumble on and was a particularly powerful stimulus for the production of a new 'Statement of general principles' issued by the parent company in March 1997, reaffirming not only the commitment to the environment but also, for the first time, respect for human rights. Clearly, something very significant was happening in the very highest echelons of the Shell family.

A year later these internal upheavals were reflected in two amazing documents, *The Shell Report – 1998* from Royal/Dutch Shell titled 'Profits and principles – does there have to be a choice?', and Shell UK's *Report to Society* (1998). Both lavishly produced 'glossies', they are aimed at the public as a whole, not just Shell shareholders – and they hardly ever mention the word profit. Instead, they unashamedly address what is now clearly seen within the organisation as an absolute requirement – public approval for their record on what might be called the social responsibilities of big business, that is towards the environment, health and safety, human rights and politics in general. Although the documents claim, probably justifiably, that plans for major changes in the overall way Shell operates, the so-called 'transformation', was begun in 1994, they openly admit that Nigeria and Brent Spar have had big influences on the way the changes have been handled. From the traditional approach of 'trust us' (in other words: leave it to us, Shell knows best) they claim now to be operating in 'tell us' mode. So they are hard-selling their green and politically correct credentials[1] and soliciting comment and advice on their past performance, their published attitudes and their future proposed actions from absolutely anyone; the documents even include prepaid postal comment questionnaires that will supposedly go straight to the top men. This is a sort of Brent Spar dialogue process broadened out to cover all societal issues which impinge on Shell's activities. It would, of course, be naive to think that all of Joe Public's 'advice' will be taken seriously, no matter how little he knows about Shell's commercial operations. But the very existence of the documents indicates the influence of Nigeria and the Brent Spar on a huge organisation that has ignored public opinion concerning its operations for decades.

If Shell had taken most of the Brent Spar flak, and was therefore most in need of image restoration, the UK offshore industry as a whole realised that it also had a communication problem with the decommissioning of redundant offshore structures. Accordingly, in July 1996 the Offshore Decommissioning Communication Project (ODCP) was established with the specific objective of promoting 'an informed debate on the issues surrounding decommissioning'. Backed by more than seventy oil-industry related companies, as well as the industry's professional organisations, this small London-based group of mainly experienced oil men had the task of providing clear and jargon-free information about the problems, and possible solutions, associated with decommissioning on behalf of the country as a whole. According to UKOOA's Annual Report for 1997, the ODCP effort that year had been mainly directed towards influencing the outcome of the OSPAR meetings through workshops, offshore visits and commissioning specialist studies of various aspects of different decommissioning options. The results of these efforts were not very obvious, and the OSPAR meeting in Brussels in September 1997 made no progress on offshore decommissioning, deferring a definitive resolution until 1998. In the meantime it was also announced in the UKOOA 1997 Annual Report, that while ODCP's identity would 'be maintained at least until mid-1998 when the OSPAR Ministerial meeting is scheduled', its activity level 'would be progressively run down now that its principal objectives have largely been achieved'. This could be interpreted somewhat cynically as meaning that the industry did not feel that the ODCP had been an unqualified success. Did it see failure at OSPAR staring it in the face?

But as the quotation from Heinz Rothermund at the head of this chapter implies, the repercussions from the Brent Spar were by no means restricted to the oil and gas industry, but were felt more or less throughout the commercial world. If Shell could be brought to its knees by public opinion whipped up by a single-issue pressure group, then so could any other company, and many of them probably much more easily. Many senior executives were worried.

An example of the tangible effects is the 1997 report of ACBE, the Advisory Committee on Business and the Environment (ACBE 1997). ACBE is technically an Advisory non-Departmental Public Body, in this case set up initially in 1991 by the Department of the Environment and the Department of Trade and Industry. Its role is to provide for dialogue between Government and business on environmental issues in order to help mobilise the business community in demonstrating good environmental practice and management. To do so, it presents reports and recommendations to the ministers in the two sponsoring departments and these are then promulgated to the UK business world as a whole. Following the Brent Spar, ACBE produced a section of its report headed 'Integrating the environment into business decisions: the consensus approach', consisting essentially of a series of recommendations about how companies might avoid the Brent Spar type of situation, and how the Government might help them. The basic premise was that there are sound commercial arguments for companies to move away from a 'decide-announce-defend' system (the 'trust me' mode mentioned in the new Shell Reports) towards one of genuine dialogue with 'interested parties' who could include almost anyone who thinks they have an interest in the environmental consequences of the company's activities. The ACBE Report does not refer specifically to the Brent Spar, but this was certainly a case in which Shell 'consulted' only those groups that they were more or less obliged to under the BPEO process, and this definitely did not include Greenpeace. We will look a little later at whether it would have made much difference if they had.

In the meantime, the need to broaden the dialogue process led to another recommendation in which ACBE called on the Government 'to review the flexibility which regulators and ministers have in approving decisions which require adoption of the Best Practicable Environmental Option'. ACBE believed that the definition of BPEO should remain unchanged, that is with the principal considerations being practical feasibility, environmental impact, safety and risk to the workforce, acceptability to the relevant authorities and other directly affected groups, and costs. But ACBE felt that strict adherence to the BPEO process might lead regulators and ministers (not to mention applicants like Shell) to fail to give sufficient attention to 'value systems of "interested parties" which fall outside the BPEO'. These might include 'aesthetic, amenity and symbolic factors' as they certainly did in the case of the Brent Spar.[2] 'As such values are often easier to communicate than factors such as scientific data and risk assessment, they are more commonly understood by the general public and the media, who may therefore be more sympathetic to the position taken by the "interested party"'. In the same vein, it points out that a claim that a particular action 'meets the law', national or international, may not be a sufficient defence against the arguments of objectors since the legal requirements usually fail to take these value systems into account.

ACBE therefore recommends that where potentially sensitive environmental issues are involved, companies engage in dialogue with as wide a range of 'interested parties' as possible at a very early stage. Since the values of the different interest groups may often conflict with those of the company, ACBE suggests the involvement of a third party facilitator might be helpful. While it might be advisable to keep the early discussions private, operating in a closed manner, still characteristic of many companies, may itself 'be a factor engendering strongly negative positions by some "interested parties"'.

In this context, the potential influence of the media is enormous, and a significant section of the ACBE Report is devoted to suggestions on how companies might approach this problem. The overall objective, the Report suggests, should be 'to avoid matters becoming the subject of sudden and unplanned sensational media reporting' (shades of the Brent Spar), since

> in the glare of publicity, constructive debate on potentially contentious matters becomes very difficult....In such an atmosphere it is much easier to appeal to the value systems of 'interested parties' than it is to argue detailed points of science, law or economics.

But sensational media coverage cannot always be avoided and ACBE recommends that companies plan well ahead for such an eventuality. The plans should include 'a very fast decision communication link with the ultimate decision makers' and resources 'dedicated to maintaining alignment of action throughout the business', presumably so that the right hand knows what the left hand is doing. These are clear references to the Brent Spar affair in which, for long periods, Greenpeace seemed to be in the publicity driving seat with Shell often following on behind in a fairly ineffectual damage limitation exercise. Because the public, and even Shell's own employees, were so badly informed about the issues when the Spar hit the headlines, the company already had an enormous philosophical hill to climb from the very beginning. This was made much worse when Shell personnel, albeit from parts of the organisation not directly involved with the Spar, openly voiced their disquiet at the proposal.

ACBE's recommendations are all very sensible and common-sense stuff, and with hindsight might well have avoided at least some of the problems Shell experienced with the Brent Spar. Indeed, many of them are characteristic of the dialogue process employed by Shell after the Greenpeace occupation and the company's new approach epitomised in its reports to society. Nevertheless, in view of Greenpeace's clear and implacable objection to any form of deep-sea disposal, it is difficult to believe that however careful Shell had been the first time round to operate a similar system, any consensus could have been reached other than to have discounted this option at the outset. This is not to denigrate ACBE's recommendations, but simply to point out that there will be many situations in which the aims and values of the different participants in the dialogue process are so diametrically opposed that no consensus is possible. At this point the services of an arbitrator, preferably independent of the government regulator, may be called for.

Most of the ACBE recommendations deal with fairly nebulous concepts like conflict resolution, hence the emphasis on 'dialogue', 'interested parties' and 'consensus'. The response of the government of the day, published with the ACBE Report two months before the election which removed it, is fairly anodyne, broadly accepting the suggestions and in some cases indicating that relevant measures were already under way. But the response to one recommendation, calling for more flexibility in approving decisions requiring adoption of the BPEO, is more interesting. This states that

> The Government will give careful thought to this recommendation over the coming months. The Department of the Environment may want to hold informal discussions with other interested parties on this recommendation, and would welcome a chance to explore it further in discussion with ACBE.

This sounds to us like civil-service-ese for 'No way', and may relate to another ACBE recommendation based on a section of the Report dealing with the role of science and risk assessment. The BPEO system, with all its uncertainties and problems, attempts to rationalise the decision-making process. All the possible options are supposed to be compared under the BPEO headings and hopefully a clear front runner emerges. Much of the process is based on measurements, numbers, statistics and probabilities, the stuff of 'incontrovertible' science. It would be surprising if civil servants, and for that matter politicians, in the relevant departments were happy to compromise this part of the system by making it more 'flexible', even accepting that aesthetic, moral and symbolic factors also have a part to play.

But the ACBE Report points out that the scientific/technical card is also a difficult one to play because of the need to improve the public perception of what science can and cannot do. 'Science', the Report quite correctly says, 'is an ever-evolving mass of knowledge and understanding about how nature works', so that 'much of science is not absolute'. It goes on to emphasise that risk assessment is also never absolute,

Yet the number of times the media ask, on behalf of the public, whether the scientific information 'means' that something is 'totally safe'...suggest a lack of appreciation of the concept of risk, and the contribution science makes towards assessing it.

The answer, the Report suggests, is that the Government should put much more effort into improving the public understanding of science and risk assessment.

Although it is difficult to identify exactly what measures ACBE is asking for in this area, or the Government is promising, the aim is laudable. But this is *not* the principal lesson to be learnt from the Brent Spar, at least as far as the use of science is concerned, because the adequacy or not of the scientific arguments on either side of the debate was not a significant factor in Greenpeace's success in stopping the deep-sea disposal plan. Some of the scientific so-called 'facts' were, of course, aired during the occupation; for example, statements about the likely serious impacts of the Spar on the marine environment from Greenpeace and claims that these impacts would be insignificant from Shell. But the bulk of the opposition whipped up by Greenpeace was not based on these arguments. Instead, most opponents were motivated by other beliefs: that Shell, with the connivance of the Government, had acted in a secretive and high-handed manner; that the deep-sea option had been chosen for expediency, financial or technical; that if you give such megalithic organisations an inch they will take a mile, that is the precedent argument; and that they had reneged on all sorts of promises, national and international. The scientific arguments were not properly debated until well after Shell had already pulled the plug on the deep-sea option and towed the Spar to Erfjord. And as we saw in Chapter 3, even then the debate was not based on different interpretations of scientific 'facts' (with all the provisos that go along with this word) but was confused by errors in the BPEO documents which resulted in discussions that were wrongly construed as indicating fundamental disagreements between well informed scientists. This problem is much easier to solve than the public understanding of science one identified by ACBE. How? By simply making sure that the errors are not in the documentation in the first place. And how do you do that? By ensuring that the people who produce the documents know what they are talking about, or at least that they are checked by someone who does.

Depending on your point of view, these statements may strike you as either naive or pretentious. Are we really suggesting that the documentation backing up BPEO cases is not necessarily written by experts? Well, based on the Brent Spar case, yes, we are. How else can you explain the presence of such basic errors as those dealt with in Chapter 3? At some point in the process someone put together the final BPEO and Impact Hypothesis documents and failed to realise that applying environmental data from abyssal depths to potential disposal sites at half that depth made a nonsense of many of the arguments. Whoever did this, or allowed it to happen, revealed a lack of knowledge that is regrettable to say the least. And it could have been avoided very simply by having the documents checked by experts, and preferably independent ones with no axe to grind.

Interestingly, this is exactly what has ostensibly been done in Shell's attempt to find a final solution to the Spar problem, though many people think that the dialogue process as a whole was a touch excessive. It is surely unthinkable that such an expensive, logistically complex and drawn-out process should set a precedent for future BPEO undertakings. Indeed, Shell more or less said as much in its press release announcing its final preferred option (see chapter 8). Nevertheless, some features could usefully be adopted as standard elements of the process. For instance, Det Norske Veritas were employed by Shell Expro specifically to check the technical, safety, environmental and costing aspects of the detailed proposals from the six shortlisted potential contractors, to ensure that all the relevant aspects had been covered and that the conclusions were sound. Following this exercise, the reconstituted Shepherd Group was asked to compare the new proposal with the deep-sea option purely from a scientific (i.e. environmental and engineering) point of view. If such a procedure, or something vaguely like it, had been in place when the original Brent Spar BPEO was being carried out it would have made no difference to Greenpeace's objections, but at least some of the obfuscating and sterile arguments about disagreements among scientists would have been avoided. Assuming that this peer arrangement works satisfactorily in this case, a similar process will hopefully become incorporated in all future BPEO studies. This alone would be a fitting and worthwhile outcome of the mess that surrounded the Brent Spar. But there were several other potentially beneficial effects, some tangible, some rather nebulous, that also came out of the affair.

Chapter 10

Epilogue

Was the Brent Spar a 'good' or a 'bad' thing?

In the Introduction we posed the question of whether the Brent Spar affair had been the victory for the environment and the little man that Greenpeace had claimed it was. We don't know because the jury is still out. The verdict will depend on whether the mistakes made and the lessons learned result in permanent changes in the way all sides deal with these problems. The short-term effect was disastrous, in the sense that the decision not to sink the Spar was prompted by the effects of the highly emotional and ill informed response by the public, particularly in Germany, whipped up by Greenpeace. This is not to say that the decision was wrong, but only that it was taken for the wrong reasons.

But what would have happened if neither Greenpeace nor anyone else had made a fuss about the disposal plan? Sometime in the summer of 1995 the complex, but well planned process of detonating the charges on the Spar's buoyancy tanks would have been completed, hopefully without injury to the personnel involved, and the structure would have plummeted to the bottom at the disposal site. It would have disintegrated to some greater or lesser extent either in the water column or when it reached the bottom. It would have disturbed some sediment, killed quite a number of animals, mostly very small ones, and started to release into the bottom waters of the North Feni Ridge its various components, oil, sludge, metals, PCBs and so on. Despite all this, the best qualified experts all agree that the effects would have been mainly extremely local and rather insignificant, certainly compared with many other impacts on the deep ocean or those that might have occurred as the Spar was transported back through the shallow inshore waters. But there would have been three much worse consequences.

Apart from those directly involved, few people would have been aware of the Spar's passing. The cold waters of the North Atlantic would have closed over the top of it, probably leaving a small oil slick, a few patches of rust and an assortment of oddments of floating debris. There would have been an obligation on Shell to conduct, or at least pay for, some degree of monitoring over the succeeding two or three years, ostensibly to check on the accuracy of the forecasts in the Impact Hypothesis Document which had accompanied the BPEO licence submission. But this monitoring programme would probably have been the minimum Shell considered it was required to do. It would probably have resulted in a couple of rather anodyne reports to SOAEFD which, after equally anodyne responses with one or two minor criticisms and questions, would have been filed away with masses of similar reports, never to see the light of day again.

Second, as explained in Chapter 4, we believe that the offshore oil and gas industry and, we suspect, the relevant government departments, would have considered this option as at least a strong possibility for other redundant structures. In considering these future disposals, both sides would have claimed that they were adhering to the avowed case-by-case principle, and if pressed would have made a big issue of this. But we doubt whether it would have been applied in quite the way the Shepherd Group believed it should.

But the third, and by far the most serious consequence, would have been that, having completed this first exercise with relatively little trouble, all sides would have seen no reason to change either the procedure or the way they operated it. Future applicants would have felt no compunction to involve or inform 'interest groups' other than those expressly mentioned in the regulations and guidance notes,[1] and certainly not the public at large. The errors in the documentation, whether caused by ignorance or ineptitude, would not have come to light, or at least not in the glare of publicity that highlighted them in the summer of 1995, and there would have been no great pressure to improve the situation subsequently. This book would certainly never have been written, nor, much more importantly, would the two Shepherd Group reports. And the government licensing departments, in this case the DTI and SOAEFD, would probably have closed the files on the Brent Spar happy that the job had been completed satisfactorily and that they had discharged their responsibilities properly. This is not as cynical as it might seem. We have no reason to believe that any arm of government, either collectively or individually, was intending to do anything underhand or wrong. Although, as we have tried to explain in Chapter 4, there was (and probably still is) a tendency in both the offshore industry and among civil servants, at least in the DTI, to favour the offshore disposal option, we are confident that they truly felt that this was the best solution, certainly in some circumstances. But until they had it pointed out to them rather forcefully, they probably did not realise that they might be making mistakes or that they might not have the in-house expertise to recognise this.

All this has changed, directly as a result of the Greenpeace action. Shell Expro and its parent company show every sign of having learned a difficult lesson well. The 'way forward/dialogue process' that they instigated following the capitulation seems to be a model of 'political correctness', while retaining the right and responsibility of the company to take the final decision. Of course, since Shell's final choice of disposal option is one that Greenpeace can endorse, it has not been an adequate test of the efficacy of this approach in other situations where the outcome does not coincide with the wishes of its opponents. Nevertheless, if a similar process, though probably with not quite the same ardour, is followed by Shell and other oil and gas companies in equivalent circumstances in the future, the confrontational situation that typified the Brent Spar might be avoided. If, as expected, a moratorium on marine disposal is agreed at the OSPAR meeting in July, a truly equivalent situation will never arise. But there are many other situations in which a more open dialogue between oil and gas companies and interest groups would be appropriate. In the broader community, if only a fraction of the ACBE recommendations prompted by the Brent Spar are adopted by UK companies and the Government, the whole subject of the environmental consequences and licensing of their activities will be much more transparent than in the past.

Environmental groups in general, and Greenpeace in particular, have also presumably to some extent been changed by the Brent Spar experience. Greenpeace claimed, and continues to claim, that the perceived errors it made in this campaign, particularly the mistaken estimate of the amount of oil remaining in the Spar, was not a significant factor in its anti-dumping arguments. It is correct. Moreover, the revelation of this error did not apparently damage significantly its direct support at the time. In fact, the organisation's campaign manager during the Brent Spar affair, Chris Rose, claimed that more new members joined the organisation because of this perceived 'honesty' in admitting the error than left in protest at the original misinformation. But whether Greenpeace likes it or not, many people who were 'casual' supporters over the Brent Spar tend to remember this error rather than much else about the campaign, partly because they are, from time to time, reminded of it by journalists who are a little more cynical towards Greenpeace than they perhaps were previously.[2] Certainly the UK television media are apparently less willing to accept and transmit Greenpeace images and interpretations uncritically than they were before the Brent Spar. If this means that Greenpeace will be a little more circumspect in its public statements it can surely only be a good thing for all concerned. In June 1997, Peter Melchett told one of us that, following Greenpeace's experience with the Brent Spar, it was being more open about its arguments and intentions at a much earlier stage than it had been in the past, pointing to the output in relation to its 'Sane energy' campaign as an example. Like the apparent objective of industry to be more open and upfront with its intentions, this should also reduce the likelihood of ill informed arguments in the heat of a crisis.

In a related initiative, in 1997 Greenpeace asked the organisation SustainAbility[3] to organise a 'stakeholder consultation project', ultimately called 'Beyond sparring', to consider the decommissioning and disposing of oil and gas structures. Rather unfortunately, the letter sent out to potential participants perpetuated one of the glaring errors in the original publicity by referring to 'Shell's decision to dispose of the Brent Spar oil platform in the North Sea'. But its heart was in more or less the right place.

The objective of the project was to go beyond the applauded Shell dialogue process for the Brent Spar and to deal with the broader question of redundant oil and gas structures in general. The aim would be to develop 'technical models of best practice' and 'decision-making models' for decommissioning and disposing of oil and gas structures through a 'multi-stakeholder "consensus-based" process'. Stakeholders in this case would presumably include a range of individuals and interest groups, much like the 'interested parties' in the ACBE Report, together with representatives of industry, governmental and non-governmental organisations and technical experts. After a 'formal stakeholder consultation process', an international working group would be formed to contribute to and review SustainAbility's consultation report and a final report would be widely circulated to 'interested parties' and international political fora such as OSPAR. According to *Energy Day* (26 June 1997) the idea 'met a cool response from a number of oil companies contacted'. Not surprisingly, the Oil Industry International Exploration and Production Forum (E and P Forum) and the industry's specialist Offshore Decommissioning Communication Project (ODCP) ultimately declined the invitation to participate. It is not difficult to see why. On the face of it, the initiative seems to be a laudable one, attempting to solve a complex and difficult problem by consensus rather than by confrontation. But as SustainAbility's introductory letter points out, 'it is no more democratic for Greenpeace, or any other individual group, to set decommissioning and disposal criteria itself' than the process they opposed in the case of the Brent Spar. Consequently,

> given its central role in sparking the decommissioning debate, Greenpeace recognises that it must clearly articulate a view of the way forward by listing the basic principles for options the organisation would support. These would constitute an Integrated Removal Plan for all North Sea installations, with the possible exception of concrete gravity-based structures.

The implication is obvious. No matter how strong the arguments, environmental, financial, safety, legal, or even possibly aesthetic, brought forward for some other option such as toppling or deep sea disposal in relation to specific structures, Greenpeace would be implacably opposed to it. This approach might just be acceptable if all of the consequences of the various options had been properly assessed and discussed openly in relation to the Brent Spar. But we believe that we have demonstrated that this was far from the case. To refuse even to consider the arguments for these options at the outset of a so-called 'consensus-based' process is hardly likely to seem constructive to at least some of the potential participants.

At the end of 1997, and in the face of continued lack of coopera-tion from the oil industry, the Greenpeace/SustainAbility project was terminated. But in early 1998 the European Union expressed a willingness to fund the project, with Greenpeace simply becoming one of the stakeholders. In a Greenpeace release announcing this new arrangement, the hope was expressed that with the environ-mental group no longer paying the piper, the oil industry would now agree to participate. This implied that the reluctance of the industry to join in the discussions was, as SustainAbility's John

Elkington had written (Elkington 1997), because they 'feared that the process was a trap, designed by Greenpeace to ensure that the outcome was a lose-lose affair for the industry'. But whether this is true or not, it unfortunately misses the point. If the ground rules for the project continued to dismiss any consideration other than onshore disposal it was unlikely that the industry would be willing to participate. But if the UK and Norway sign up to the possible OSPAR total ban on disposing of anything in the ocean later in 1998 (other than a few structures that would be virtually impossible to remove) there will, of course, be no legal alternative to bringing everything ashore. The Brent Spar will then have been the precedent that Greenpeace hoped it would be, and possibly not only for the disposal of offshore installations but also for other much more serious waste-disposal problems. For those who believe that the oceans just might help humankind solve some of these problems in the future, this could negate all the good things that have come out of the affair.

Finally, there were two very late additions to the Brent Spar equation, at least as far as this book is concerned as it was being prepared for the press in late June 1998. The first was the Shepherd Group Report, that is the second report of the NERC Scientific Group on Decommissioning Offshore Structures, which was released on 30 June and whose conclusions we touched on in Chapter 8; the other was the book-length Greenpeace view of the affair, *The Turning of the Spar* (Rose 1998) written by Chris Rose, the environmental group's Campaigns Director in 1995, which also appeared in June. In our opinion they belong on opposite sides of the scales.

As we have already said, the Shepherd Report is a very thorough and detailed document. Unfortunately, large sections of it are written in extremely tortuous and obfuscating language that will do little to convince the layperson that scientists can ever call a spade a spade even when it manifestly is one. But some points are very clear, and it will be difficult for Greenpeace to take much solace from the report, even though, as we have also said, it should allow the government to license the disposal option for the Spar which Greenpeace can accept.

Although the report is quite severely critical of some aspects of Shell's dialogue and 'way forward' programmes, and particularly the way some of the data were presented and interpreted, it nevertheless suggests that the exercise 'probably represents one of the most comprehensive and thorough BPEO studies ever undertaken'. In turn, the group's own deliberations must also be one of the most detailed analyses of such an exercise ever conducted. And they found the system wanting.

Large sections of this book have dealt with the way the BPEO system was misused, albeit inadvertently, in the original studies on the Brent Spar, resulting in a great deal of confusing information. Much of this resulted simply from a failure to have relevant specialists check some of the basic scientific and technical information at a sufficiently early stage. The Shepherd Group itself effectively acted in this capacity in the second iteration, pointing out areas needing clarification or further work. But in doing so they identified more subtle, yet fundamental, problems in the BPEO process, particularly in dealing with extraordinary situations – which the Brent Spar certainly was.

As the Group point out, the Royal Commission's intention in establishing the BPEO system was that it should enable, in any problem to which it was applied, the selection of the environmentally most acceptable option from a range of technically feasible and acceptably safe possibilities. The corollary is that it is no use choosing an environmentally acceptable option that is technically unachievable. But where, as in the case of the Brent Spar, some of the technical tasks associated with the various options are very complex, it is difficult to assess all the implications for health, safety, the environment and even cost without an extremely detailed analysis. Such analyses are not normally undertaken until a relatively late stage in the process. Indeed, the Shepherd Group felt that such uncertainties still remained in the documentation for Shell's preferred option even when they examined it. They therefore suggest that the BPEO system should include a peer-review process, presumably at an early stage, to identify those particularly important feasibility issues that should receive detailed study. This is particularly important where, as in the case of the Brent Spar, the environmental impacts of several of the options are both small and insufficient to distinguish between. The implication is that it would be pointless to agonise about small environmental impact differences when the differences in other BPEO criteria might be much more significant. Similarly, the emphasis in the BPEO guidelines on comparison between the options, particularly in the environmental area, comes in for criticism. For example, the Group felt that the energy requirements, and the calculated emissions of gases to the atmosphere, of several of the options were very small relative to many other activities so that 'there was little justification for this assuming the role of an important selection criteria'. The overall impression is that it believed that there is a danger of the BPEO system encouraging concentration on trivial detail, particularly if there is an over-emphasis on environmental issues to the exclusion of potentially more crucial or difficult factors. Following on from this, although the report does not say so, or even vaguely imply it, it would be tragic if lives were lost in turning the Brent Spar into a quay extension simply to satisfy the principle of not causing any impact on an already less-than-pristine ocean.

As we have already said, we are confident that notwithstanding this second Shepherd Group Report, the Government will eventually issue Shell with the licence to proceed with the Wood-GMC disposal option. The DTI and the DoE will find some parts of the report difficult to swallow. So will Greenpeace, particularly those which suggest there is little to choose between the onshore or deep-sea disposal options on environmental impact grounds alone. But as with the Group's first report, this one again emphasises the over-riding importance of public acceptability, the field that Greenpeace play particularly well on and which is largely the subject of Chris Rose's book.

The book originally appeared on the internet and as manuscript advance copies in late January 1998 under the title *Consequences of the Brent Spar Victory*, a few days before Shell's announcement of their latest preferred disposal option for the Spar. It had apparently been written in a great rush and, possibly as a result, contained a number of blatantly erroneous statements of apparent fact. Except for the people they directly affect, these are not terribly important, though they could easily have been removed from the published version – but were not.[4]

Coming as it does from Greenpeace, it is understandably as biased and self-congratulatory as you might expect from the other side. And why not? After all, as Rose says in his introductory sentence, 'The Brent Spar campaign is the single most obvious Greenpeace success of recent years', so you would expect them to milk it as much as they can. Although it contains some unnecessarily personal attacks on a number of individuals, it is generally a well written and clear account of what happened, when and why. But rather than trying to identify the errors that were made on all sides, and thereby improve the decision-making process, Rose has a number of clear objectives that prevent him from dealing with the facts and events even-handedly. First, he seeks to justify the Greenpeace anti-dumping campaign against what he sees as a background of UK Government and offshore industry complicity. Second, he emphasises the perceived errors in Shell's scientific case for the Spar's deep-sea disposal, and downplays the Greenpeace statements and actions that drew criticism from varied quarters at the time and subsequently. Third, he identifies what he sees as the effects of the Brent Spar 'victory' on industry (particularly, of course, the oil industry), on the media, on the UK Government and, through OSPAR, on Europe.

In each of these areas there is a good deal in what he says. To take the last one first, Greenpeace can certainly take a good deal of the credit for the major changes in British industry in general which we outlined in Chapter 9, and when he wrote the book he hadn't even seen the latest Shell reports!

He is also certainly correct in saying that, prior to the Brent Spar, Shell had been in close touch with the Department of Trade and Industry in the lead-up to the proposed deep-sea disposal; and that for both the offshore industry and the DTI this option was the preferred one because it was the simplest and cheapest and would prepare the ground for possible future deep-sea disposals. Similarly, as he says, the cheapest option would also appeal to the Treasury because the tax relief that could be claimed on decommissioning costs would have meant less tax relief and therefore greater income to the government. This at least partly explains the UK Government's anxiety, along with that of Norway, to keep the *in situ* and deep-sea disposal options open in the OSPAR discussions.

He is similarly correct in pointing out the factual errors in Shell's BPEO documentation, the same ones that we have emphasised. But he does not admit explicitly that these were in effect irrelevant, since even if they had not been present it would have made no difference to Greenpeace's objection to the deep-sea disposal. Indeed, while he refers extensively to the statements in the first Shepherd Group Report where these can be construed as supporting the Greenpeace case, he does not point out that the group also said that a dumped Brent Spar would not have caused a significant environmental impact notwithstanding the errors in the BPEO documents. Nor does he point out that Dr John Gage, whose criticisms of the documents are used to support the Greenpeace 'scientific case' against the dumping, was a signatory not only of the first Shepherd Group

Report, but also the second one which concludes that the deep-sea impact of the Spar would have been even less than they first thought.

On the other hand, he is at great pains to play down the significance of the erroneous Greenpeace claims about the Spar's contents. Like us, he believes that their impact on public opinion at the height of the furore was minimal, receiving very little media coverage until much later. Even when they did receive attention with Greenpeace's admission of the 'oil' error and DNV reporting that they could find no evidence of either the oil or the 'concrete eggs', it apparently made little difference to the public conception of Greenpeace, which tended to be praised for its honesty rather than castigated for its mistakes.

All in all, the book reads like a catalogue of 'we were right and they were wrong', with never a suggestion of 'OK, the science may not support us, but morally we are unassailable because it is simply not right to use the oceans as a dump'. There is little indication that Greenpeace learned any philosophical lessons from the Brent Spar.[5] This is a shame. There is a great deal of respect for, and empathy with, environmental groups like Greenpeace, even among those who find themselves the target of their activities. Even when their science and rationality are weak, their moral arguments are often extremely powerful.

We have argued consistently in this book that the technical and scientific information used in sensitive environmental decisions should be of the highest possible quality. But at the end of the day, the public is likely to back the argument that seems to be based on concepts such as fair play, good versus evil, right versus wrong, the little man versus the power of the establishment. And why not? We are an irrational, emotional species. This is what makes us so fascinating. The onshore fate of the Brent Spar, essentially brought about by public protest stimulated by Greenpeace, may well be the 'right' one. But when we take these decisions, individually or collectively, we should not kid ourselves, nor let anyone else kid us, that they are based clearly and unequivocally on cold, dispassionate rationality. They certainly were not in the case of the Brent Spar.

International law and agreements: did the UK Government have the right to issue a licence for the deep-sea disposal of the Brent Spar?

Throughout the debate about the pros and cons of the deep-sea disposal of the Brent Spar, and quite apart from the arguments about whether or not it would cause significant environmental damage, there were repeated references to the UK's international agreements and its obligations under a variety of conventions. Greenpeace and its supporters maintained that in granting a licence for the Spar's deep-sea disposal the Government had at best reneged on various promises made to other countries, including our European neighbours. Worse, they claimed, the Government had broken various firm international agreements or even the International Maritime Law. The Government (and Shell) on the other hand, consistently claimed that they had acted, and would continue to act, strictly in accordance with the law and the conventions to which the UK was a party. Let us see if we can get somewhere near the truth.

The international 'rules' governing what the UK, or for that matter any other European nation, can and cannot do in the seas surrounding their coasts are determined essentially by agreements made within two organisations, the United Nations Conference on the Law of the Sea (UNCLOS) which deals with maritime law in general, and what is now known as the Oslo-Paris or OSPAR Convention which is concerned specifically with waste disposal in the northeast Atlantic.

The first UNCLOS conference met at Geneva in 1958 to attempt to regularise a rather confused and informal set of arrangements governing the use of the seas which had developed somewhat erratically over the previous several hundreds of years. Apart from the rights of 'innocent passage', most maritime nations at that time claimed jurisdiction over a three-mile-wide strip of water bordering their shorelines. But beyond these 'territorial waters' the seas were largely considered to belong to everyone and therefore available to any nation to do virtually anything. However, there had always been arguments about the precise extent of territorial waters and the maritime states' rights over them, and the rights of access to the sea's resources, particularly fish.

UNCLOS I failed to reach agreement on the important issues of fishing and conservation of living resources, but it did make considerable progress on other areas, including those relevant to the Brent Spar. The resulting agreement, usually referred to somewhat imprecisely as the Geneva Convention, came into force in 1964 after one-third of the participating nations had ratified it. It formalised the old idea of a three-mile-wide territorial sea over which the coastal state had more or less total control, but also introduced the idea of a second belt, a nine-mile-wide 'contiguous zone', outside the territorial sea and over which the adjacent state had less, but still significant control. More importantly, at least from the point of view of the Brent Spar, the Geneva Convention included a number of statements about the rights and obligations of nations to exploit the seabed resources on their adjacent continental shelves, roughly the shallow areas over the seaward extension of the landmass before reaching the deep sea proper.

The first formal claims to such rights had been made in the 1940s, with the USA claiming unqualified sovereign rights to its continental shelf resources in 1945. By the time of the Geneva Convention some twenty nations had claimed some form of jurisdiction over their shelves and the Convention confirmed these rights, but only for the purposes of exploring and exploiting the natural resources, including oil and gas. But it categorically stated that abandoned or disused installations resulting from these activities should be entirely removed from the sea. This was fairly unequivocal and the UK not only accepted the ruling but actually proposed it (see Gao 1997). This is not surprising since few such installations existed at the time, and none in UK waters. Twenty years later the situation had changed dramatically.

Because the Geneva Convention had failed to resolve the problems of fishing and other living resource exploitation, a second UNCLOS meeting was convened in 1960, that is even before UNCLOS I came into operation. Like the previous one, UNCLOS II also failed to reach agreement, and over the next decade or so a series of unilateral claims to fishing limits, or even territorial waters, extending up to 200 miles offshore made the situation even more

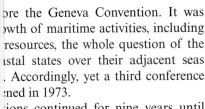

ɔre the Geneva Convention. It was
ɔwth of maritime activities, including
resources, the whole question of the
ıstal states over their adjacent seas
. Accordingly, yet a third conference
ːned in 1973.

ːions continued for nine years until
d document for signature was finally
· eleven years to obtain the signatures
ː hundred and fifty-nine participating
he sixtieth signature, that of Guyana,
ȝ and the convention came into oper-
theoretically only for the ratifying
was not one of them and had not
the time of the Brent Spar. It finally
change of government the previous

and arguments during the long
ncerned the thorny problems of the
ıs and how to measure them from
ːe definition of Legal Continental
ment also introduced yet another
ːe Economic Zone or EEZ, extending
ᴢ00 miles from the coastal baseline and within which the coastal
state had 'sovereign rights to conserve, manage, explore and exploit'
all living and non-living resources in the water column and in and
under the seabed. By this time a number of nations, including the
UK, were actively engaged in offshore oil and gas exploitation, and
the problems of the disposal of the resulting redundant installations
was becoming an important issue. At a late stage in the negotiations
the UK proposed the following addition to the final document
which, after a full debate, was eventually adopted as part of Article
60(3) (see Side *et al.* 1993):

Any installations or structures which are abandoned or disused
shall be removed to ensure safety of navigation, taking into
account any generally accepted international standards established in this regard by the competent international organisation.
Such removal shall also have due regard to fishing, the protection
of the marine environment and the rights and duties of other
states. Appropriate publicity shall be given to the depth, position

and dimensions of any installations or structures not entirely removed.

This is clearly at variance with the Geneva Convention, since it contemplates less than the total removal of some structures in appropriate cases. Even before UNCLOS III came into force, and certainly before the UK acceded to it, the approach of the British Government, and of several other states with similar interests, was that this was the *de facto* internationally agreed rule and superseded the Geneva Convention which was technically still in force. The legal niceties of this curious situation in which the UK Government disregarded an old rule to which it was a signatory, while instituting a new one to which it wasn't, is discussed by Gao (1997), who concludes that clarification by early ratification of the 1982 Convention would be highly desirable, to say the least. Although this ratification has now taken place, we assume that even before this the Government's lawyers felt that they were on firm ground. In any case, the UNCLOS III statement was still rather woolly. So now let us look at it in a little more detail, and particularly the phrase 'generally accepted standards established by the competent organisation'.

This phrase was clearly directed at the International Maritime Organisation, an arm of the United Nations, which accepted the role as the 'competent organisation' and undertook the task of drafting standards on installation abandonment. The result was IMO Resolution A.672(16) issued in 1989 and accepted as 'customary international law' at the end of 1995. Its objective was mainly to safeguard navigation, but also fishing activities and the marine environment in general. Its strongest, and most unequivocal recommendation, subsequently accepted by all relevant European parties including the UK, was that from 1 January 1998 'no installation or structure should be placed on any continental shelf or in any exclusive economic zone unless the design and construction of the installation is such that entire removal upon abandonment or permanent disuse would be feasible'. But in the context of the Brent Spar it is the resolution's statements on what should be done with existing structures that are relevant. Although it asserts that all abandoned structures should be removed completely where feasible, the resolution allows for partial removal or even non-removal in compliance with certain guidelines. These again address mainly the navigational and safety issues, but they also establish standards for the partial removal of structures in the following paragraphs:

3.1 All abandoned or disused installations or structures in less than 75 metres of water and weighing less than 4,000 tonnes in air, excluding the deck and superstructure, should be entirely removed.

3.3 Any abandoned or disused installation or structure, or part thereof, which projects above the surface of the sea should be adequately maintained to prevent structural failure. In cases of partial removal...an unobstructed water column sufficient to ensure safety of navigation, but not less than 55 metres, should be provided above any partially removed installation or structure which does not project above the surface of the sea.

The guidelines also make vague reference to the fact that no significant environmental damage should be caused by the decommissioning process and that 'potential for pollution or contamination of the site by residual products from, or deterioration of, the offshore installation or structure' should be taken into account. But the guidelines clearly refer specifically to structures standing on the seafloor. They make no mention, one way or the other, of the practice of towing decommissioned structures away from their work site to be scuttled in deep water. In the absence of any clear condemnation of such actions, the UK Government assumed, justifiably from their point of view, that the IMO guidelines gave tacit approval to license the deep-sea disposal of offshore structures.

But the UK is a signatory to another series of relevant international conventions, those concerned with marine pollution which originated from the so-called 'London Dumping Convention' of 1972, more correctly the London Convention on Prevention of Marine Pollution by Dumping of Wastes and Other Matter. This dealt with the dumping of all sorts of wastes at sea and was supplemented by a series of other conventions dealing with specific wastes or particular maritime regions. Wastes originating from the exploration and exploitation of seabed mineral resources in the northeast Atlantic were covered by the 1972 Oslo Convention with additional guidelines issued in 1991 'on a trial basis'. This convention and guidelines prohibited the marine disposal of certain particularly toxic substances such as biocides and PCBs except in trace amounts, while it required that bulky wastes, such as the Brent Spar, should be placed in waters more than 2,000 metres deep and at least 150 nautical miles from the nearest land. It also recommended that an Impact Hypothesis should be developed to predict the consequences of any marine disposal and that operators should conduct baseline surveys prior to disposal and clearly define the objectives of any subsequent monitoring activities. Finally, the guidelines stated that 'Contracting parties intending to dispose of offshore installations at sea should notify the other Contracting Parties by providing the information requested and the reasons for disposing of the offshore installation at sea'. This is why the UK Government informed its European partners in February 1995 of its intention to issue a licence for the disposal of the Brent Spar and, in the absence of any response within sixty days, assumed that none of these partners had any objection to the plan.

The provisions of the Oslo Convention were adopted in 1992 by what became known as the Oslo-Paris or OSPAR Convention which was intended to supersede and replace the earlier conventions. The OSPAR Convention also required that 'permits for disposing of offshore installations and pipelines' be issued on a case-by-case basis, and that all substances likely to 'result in hazards to human health, harm to living resources and marine ecosystems, damage to amenities or interference with other legitimate users of the sea' be removed (apart from trace amounts) before a permit is issued (House of Lords Select Committee on Science and Technology 1996: 14).

The UNCLOS III regulations, the International Maritime Organisation resolution and the OSPAR Convention in force in 1995 definitely did not encourage contracting parties to dispose of structures at sea. Furthermore, the 1991 OSPAR guidelines included the recommendation that, at the very least, disposal at sea must be shown to be the option of least detriment. But they equally clearly did not preclude deep-sea disposal of wastes under some circumstances. Whatever happens to international maritime law in the future, or indeed may have happened before this book sees the light of day, this was the situation at the time of the initial Brent Spar controversy. However much individuals, groups of individuals or even nations dislike the idea, the British Government had every right, at least on their reading of the international agreements, to issue a licence for the disposal of the Brent Spar, and its contents as specified by Shell Expro, at the selected deep-sea site.[1]

But the situation is not, of course, static. For example, at a meeting of the Oslo and Paris Commissions in Brussels in June 1996, eleven of the member states supported a total moratorium on the disposal at sea of decommissioned offshore structures, even though they had all been signatories to the original OSPAR convention which did allow such disposal for structures standing in more than 75 metres of water and as long as they were considered on a case-by-case basis. This moratorium was opposed, and not signed, by the UK and Norway, the only OSPAR members with offshore installations in such deep water. The entire subject of decommissioning was due to be debated again within the OSPAR Commission in September 1997, but this meeting was ultimately postponed until July 1998. One proposal for discussion at the meeting will be that the moratorium should be strengthened into a total ban (and see note 1, chapter 2, p. 167).

With the change of government in May 1997, the future attitude of the UK towards deep-sea dumping became particularly unclear. Three weeks after the election, the Department of the Environment informed a Heads of Delegation meeting of the OSPAR Commission in London that it would be reviewing its approach to dumping at sea. A DoE spokeswoman was quoted by *Energy Day* (22 May 1997) as saying that current policies would continue for the time being, but the option was being left open for new ministers to review the existing position. A similar statement also came from the DTI.

On 2 September 1997, to coincide with a meeting of the OSPAR Commission in Brussels, Michael Meacher, the Minister for the Environment, announced a new UK attitude towards marine waste disposal, including 'a general presumption against sea disposal' of offshore oil and gas installations. However, the 'new' British position carried the important caveat that onshore disposal was to be instituted 'wherever it is safe and practicable to do so' (letter from Michael Meacher to Llew Smith MP). This, and Mr Meacher's statements in broadcast interviews, was widely interpreted as referring to the *in situ* abandonment of parts of some of the very large concrete-based structures that would be extremely difficult to remove. But the Minister's letter went on to say that 'we continue to believe that we need to be able to examine matters case-by-case…and to deal with the occasional exceptional case, such as where installations have been damaged'. (The Brent Spar, remember, was quite severely damaged.)

The new government position was broadly welcomed by Greenpeace and the press, with Greenpeace maintaining that the announcement 'totally vindicates Greenpeace's campaign against dumping the Brent Spar in the ocean. The UK Government has accepted that redundant oil installations should as a rule be disposed of on land' (press release, 2 September 1997). Some of the press commentators were rather less impressed, implying that some of the new promises were either unnecessary or impractical. *The Times*, for example, carried an editorial on 3 September suggesting that

> In the case of abandoned oil rigs, the equation is very finely balanced, and the expectation remains that some 30 or so of the bigger platforms, or those that are damaged, will indeed be dumped at sea. The fact that the UK Offshore Operators Association accepts this with apparent equanimity means that it represents less of a shift than Mr Meacher would like us to believe.

That this was so became clear a year later, in the lead-up to the OSPAR meeting in Lisbon planned for July 1998. A leaked document to the preparatory OSPAR Heads of Delegation meeting in London in early June from Alan Simcock, a British civil servant and Acting Chairman of the OSPAR Commission, suggested that the UK, at least, would be anxious to maintain the case-by-case approach for installation disposal and that the possibility for marine disposal for some parts of some large structures should remain an option. Since, as pointed out above, the total removal of the vast majority of North Sea structures is already agreed, the document pointed out that this option could apply at most to about 122 installations (about half each, incidentally, in the UK and Norwegian sectors). This was immediately seized on by Greenpeace and much of the press as a firm intention by the UK to dump 120 rigs in the North Sea, and that Mr Meacher had been somewhat premature in his suggestion that the new Government's position was significantly different from its predecessor. By the end of the preparatory meeting it was being suggested that a compromise proposal from Denmark was likely to be accepted at the Lisbon meeting. This was expected (see *Upstream*, 11 June 1998) to call for a general ban on marine dumping except for some twenty-five concrete and 10–20 deep-water steel platforms for which onshore disposal posed particularly difficult technical problems. If adopted, this would indeed be a significant tightening of the internationally agreed rules, and very different from the situation that existed at the time of the Brent Spar.

The long list

The following 'long list' of potential contractors, and outlines of their suggested disposal solutions for the Brent Spar, was released by Shell on 14 August 1996. Those marked with an asterisk were eventually to form the shortlist from which Shell selected the final option.

*Wood-GMC (a joint venture company formed from John Wood Group plc of Aberdeen and Maritime GMC of Stavanger, Norway)

Re-use as a new harbour extension at Mekjarvik, Norway.

1   Remove topsides, possibly for re-use as an onshore training facility.
2   Raise main body with jacks on barges adjacent to Spar.
3   Cut body horizontally into sections c.12 metres thick for transport to Mekjarvik and use as harbour extension.

*Brown and Root Energy Services, London

Onshore recycling.

1   Up-end Spar to horizontal position.
2   Tow to Nigg, Scotland.
3   Recycle and re-use onshore.

Rhodes Offshore, Toronto, Canada

Onshore dismantling and recycling.

1 Tow Spar vertically to Newfoundland.
2 Dismantle vertically.
3 Transport sections onshore for cleaning, further dismantling and total recycling.

Hollandsche Staalbouw Maatschappij BV, the Netherlands

Recycle topsides, re-use tanks for onshore oil storage.

1 Remove topsides and transport onshore for recycling.
2 Raise main body vertically using a hoisting frame on pontoons.
3 Cut column into nine similar sections for transport ashore and conversion into land storage tanks for oil.

*Kvaerner Stolt Seaway Alliance (Kvaerner Installation AS, Norway; Stolt Comex Seaway Ltd; Seaway Heavy Lifting Ltd)

Re-use and recycling.

1 Tow Spar to Hanøytangen, near Bergen.
2 Remove topsides for re-use as shore-based training centre.
3 Recycle steel of main body (technique not specified).
4 Re-use ballast as rock fill.

Taylor Woodrow Construction Ltd, Southall

Onshore dismantling for recycling or re-use.

1 Repair damaged tanks *in situ*.
2 Up-end to horizontal using compressed air in deballasted tanks to equalise internal and external pressures.
3 Tow to UK dock for further operations.

ROS Holland BV, Ijmuiden, the Netherlands

Total onshore recycling, or partial recycling and re-use of topside as offshore training centre.

1   Upend to horizontal by pressurising and pumping tanks.
2   Tow horizontally to demolition site near Amsterdam.
3   Recycle and re-use (techniques not specified).

*AMEC Process and Energy, Aberdeen

(a) Re-use as rock-filled artificial reef or causeway and marina.

   1   Vertical dismantling in sections (technique not specified).
   2   Transport cleaned sections to re-use site.

(b) Recycle ashore.

   1   Upend to horizontal (technique not specified).
   2   Tow to shore facility on Teesside.
   3   Dismantle and recycle.

Mayer Parry Recycling Ltd, Erith

Onshore dismantling and recycling.

1   Partial deballasting in Erfjord to 87m draught.
2   Tow to within six miles of dry dock at Loch Kishorn, Scotland.
3   Up-end to horizontal position.
4   Move to dry dock for cleaning, dismantling and recycling.

Land and Marine Engineering Ltd, Bromborough

Onshore dismantling and recycling.

1   Remove parts of deck structure by crane-barge in Erfjord.
2   Tow vertically close to Loch Kishorn, Scotland.
3   Partial deballast to allow access to Kishorn.
4   Deballast vertically.
5   Freeboard sections cut, crane lifted and barge transported to land.

*Thyssen Stahlunion GMBH, Dusseldorf, Germany, in conjunction with CETEM, Germany, Rogalund Consultants, Norway, and Aker, Norway

(a) Vertical dismantling and recycling.

1   Topsides removed at Erfjord
2   Tow to Aker NC yard near Stavanger.
3   Attach to barges and provide additional buoyancy to lift Spar.
4   Move to location where base is grounded.
5   Cut off freeboard section and remove to shore.
6   Repeat until Spar totally removed.

(b) Re-use as water desalination plant.

1   Repair damaged tanks and clean in Erfjord.
2   Modify topsides into wind-powered generator and desalination plant.
3   Move to unspecified deep locality in Norway to provide fresh water.

UMOE Haugesund AS, Norway

Onshore dismantling.

1   Remove topsides in Erfjord by semi-submersible crane vessel (SSCV).
2   Up-end lower part to horizontal by deballasting and using SSCV.
3   Insert barge beneath horizontal Spar.
4   Tow to graving dock in Haugesund.
5   Clean and dismantle in dry dock.

Jan De Nul NV, Belgium

Shallow-water burial.

1   Removal of topsides for onshore recycling or re-use.
2   Tow Spar vertically to disposal site in about 100m of water.
3   Dredge 30m-deep trench and stockpile spoil one mile away.
4   Sink Spar into trench and cover with spoil.

Ramboll/Monberg & Thorsen Consortia, Esbjerg, Denmark

(a) Offshore re-use.

    1   Cut Spar in two.

    2   Upper part repositioned in *c.*60m water depth as accomodation or drilling platform.

    3   Lower part brought ashore for scrapping.

(b) Onshore scrapping.

    1   Remove topsides and bring ashore.

    2   Reverse-up-end remainder, possibly using polystyrene balls for buoyancy.

    3   Bring ashore on barge.

*McAlpine Doris JV, Hemel Hempstead

Onshore scrapping with re-use in quay extension.

1 Transport Spar to deep-water UK site.
2 Reverse-up-end using compressed gas.
3 Transport horizontally to UK dry dock.
4 Onshore dismantling, but possible re-use of hull sections in quay extension.

NNC/Cammell-Laird, Knutsford

Onshore scrapping.

1 Reverse-up-end with lifting equipment and buoyancy aids.
2 Repair ruptured tanks before transporting horizontally to Birkenhead.
3 Dismantle and scrap.

Heeremac v.o.f., Leiden, the Netherlands

Onshore scrapping or refurbishment.

1 Remove topsides and transport ashore for dismantling/refurbishment.

2   Reverse-up-end using either one or two heavy lifting vessels or raise vertically and cut into sections.

3   Transport to shore, but no details of final fate.

AKER Offshore Partner AS, Stavanger, Norway

1   Move in vertical orientation to deep-water site at Stord, Norway.

2   Raise vertically and cut into sections (no details).

Rhodes Offshore Partners Inc, Toronto, Canada

Onshore recycling in Canada.

1   Tow in vertical orientation to Bull Arm, Newfoundland.

2   Raise vertically and cut into sections (no details).

3   Recycle ashore.

Hollandsche Staalbouw Maatschappij AR, Gouda, the Netherlands

Re-use as oil storage tanks.

1   Remove topsides and dismantle ashore.

2   Raise Spar vertically and cut into nine cylindrical sections.

3   Sections refurbished, fitted with floor and roof and placed on prepared foundations on shore.

Hollandia BV, AA Krimpen a/d, Ijssel, the Netherlands

Re-use for wind- and wave-power electricity generation.

1   Remove topsides and fit with three windmills.

2   Tow main body to deep-water site off Scottish coast to be reunited with topsides.

3   Fit wave-power units and connect to grid.

Notes

Introduction

1 Two years later, Esso's parent company Exxon had taken over Shell's role as Greenpeace's public enemy number one. By this time it looked as though Shell would dispose of the Brent Spar onshore, as Greenpeace wished, while Shell UK's chief executive had publicly admitted concern over global warming and the role played by the burning of fossil fuels. In contrast, Exxon's US chief executive, Lee Raymond, was a leading figure in the Global Climate Coalition, the interestingly named US oil and gas companies pressure group urging President Clinton not to accede to further emission limits resulting from the Kyoto climate change summit. In the build up to Kyoto, Esso's London headquarters attracted Greenpeace protest groups while Shell's offices were left alone.

2 There were dozens, if not hundreds of references in the media to the intention to dump the Spar in the North Sea. These erroneous statements came from journalists, members of the public, politicians including government ministers, and from Greenpeace spokespersons. Although Greenpeace never intimated in its official press releases that the Spar was to be dumped in the North Sea, it undoubtedly contributed to the impression that it was – not least by the 'Save the North Sea' banner draped across the Spar's heli-deck during the occupation and images of which were flashed across the world, and major advertisements which appeared in both the *Guardian* and the *Independent* on June 8 1995 – clearly referring to the proposed dumpsite as in the North Sea.

2 Decommissioning and the BPEO process

1 As this book went to press in June 1998, it looked increasingly likely that the OSPAR meeting due to take place in Lisbon from 20–24 July 1998 would agree a more or less total moratorium on the marine disposal of North Sea offshore structures with the exception of a very small number including perhaps 20–25 concrete-based installations. Deputy Prime Minister, John Prescott, had played a key role in the

Kyoto climate summit in 1997 and was believed to be anxious that the UK's resulting 'green credentials' should not be damaged at OSPAR (see *Upstream*, 12 June 1998). In fact, the meeting went further. All structures weighing up to 10,000 tonnes must be removed completely, as must the topsides of all larger structures. Even the footings of these larger structures must be removed if at all feasible. So only the concrete bases of some of the largest installations would be allowed to stay in the sea. By implication, the intentional dumping of anything in the deep ocean would be totally prohibited.

2 The Environment Agency issued new guidelines for applicants for licences under the BPEO process in 1997, but these were not in place, of course, at the time of the original Brent Spar exercise.

3 The scientific debate

1 With no guidance on potential dumpsites, Metocean, not unreasonably, considered sites that had already been used for the disposal of waste in the past. In so doing they touched on a very murky episode in recent British maritime history. From the 1950s to the 1980s, prior to the present much more stringent national and international controls on waste dumping in the ocean and the establishment of EEZs, a range of unwanted industrial wastes were simply thrown into the sea under licence from the UK Government at a number of localities in the North Atlantic. During the same period the military also scuttled a number of ships containing unwanted chemical warfare munitions in depths ranging from 500 to more than 4,000 metres (see, for instance, ECOS 1996).

The most publicised dump site was a sort of international one, some 1,000km southwest of Land's End, used by several European nations to receive low-level radioactive waste more or less annually until it was stopped in 1983 largely as a result of Greenpeace protests.

But two sites looked at by Metocean (called UK/1 and Ukd) along with about half a dozen others were specifically used for UK waste. It is quite difficult to find out exactly what was dumped at these sites, but UK/1, some 700km west of the Scillies, certainly received considerable numbers of drums of waste material, including several tonnes of arsenic, during the early 1980s. As recently as 1982 it also received a complete ship when the MV *Essi-Kari* was scuttled there. The vessel had previously belonged to Associated Octel, a firm handling and transporting various chemicals including highly toxic tributyl lead, used until recently as a petrol additive. When one of us (ALR) contacted Associated Octel to find out more details of the vessel and its contents they were unable to find the documents because of a recent office reorganisation.

But the UK was not, and is not, the only nation to dump things in the sea. In January 1995, for instance, the Portuguese government informed the OSPAR Commission that on 24 October 1994, six months before the Brent Spar was occupied, it had dumped a ship loaded with more than 2,100 tonnes of redundant munitions in the Atlantic some 215 nautical miles from its coast at a depth of 4,000m. Although, according to the report to the Commission's Working Group on Sea-based Activities (SEBA), 'despite some media reports to the contrary, [the operation] had gone smoothly. A small explosion had occurred at the time of the dumping due to some flares being displaced; this was however of no significance'. There is no evidence in any of the reports that Greenpeace made any protest.

2 The Gage and Gordon letter to Dr Helen Wallace of Greenpeace was eventually published on the Greenpeace International website as part of the group's protest at the 'Scare stories' documentaries screened by BBC2 in late 1997, and particularly the programme on waste disposal screened on December 11.

3 The 'worm' story. This is an amusing, but rather worrying, example of how scientific statements can become garbled. A full account appeared in *Ocean Challenge* in 1996; the following is a brief resumé.

In late May 1995, at the height of the Brent Spar debacle and with the Spar still in the wilds of the northern North Sea, *Daily Telegraph* environment journalist Charles Clover phoned me (ALR) because, he said, he understood that I believed dumping the Brent Spar would do no harm. I said that was nonsense. What I had actually said was that of the two alternatives then being considered, I believed the deep-sea disposal carried less risk of serious environmental damage than dragging it back inshore. So he asked what I thought would be the impact of the Spar dumped in deep water. I suggested that it would have a number of effects and that, for a start, when it hit the bottom it would kill millions of animals, mostly very small and mainly nematode worms, living in the large quantity of mud that it would disturb. (As it turns out, I actually overestimated the effect because the dumpsite has a much firmer seafloor than I, or anyone else, thought at the time; so less mud would be disturbed, and less animals killed, than I had thought).

I then pointed out that nematode worms, though different species of course, live in soil, but that no-one seems to be concerned that millions of them are killed each time we build a section of motorway, or even a large hospital.

On 31 May, Clover referred to this conversation in a piece about the Brent Spar in the *Telegraph*. He slightly misquoted me as saying 'Some animals [not millions] – mostly worms and bivalves – will be killed. But people don't seem to ask how many millions of creatures – worms and so on – are killed when you build a motorway or a hospital'.

Nobody apparently picked this up. Then, on 20 June, as the Spar approached the dumpsite, Clover again used our conversation, this time in a piece headed 'Storm in Atlantic teacup'. This time he had me saying that a dumped Spar would kill a *number of worms*, equivalent to the number killed by the building of a hospital or a quarter of a mile of motorway.

Two days later Shell capitulated and the Major–Redwood battle drove the Spar from the headlines. On 30 June *Private Eye* referred to the *Telegraph* article, having me now say that the most likely impact of rig-dumping would be the death of *some* worms – equivalent in number to worms killed by the building of a hospital or road.

By now Tim Eggar had become President of the Board of Trade and, on 12 July, faced hostile questioning in the House of Commons about the Government's handling of the Brent Spar issue. According to *Hansard*, Tony Banks asked if the Minister could 'assure the House that environmental considerations are uppermost in his mind and that dumping something at sea is more environmentally advantageous than bringing it ashore?' In his reply, Mr Eggar said

> as for the environmental damage…I shall quote Dr Tony Rice, an independent scientist and probably the leading independent deep sea biologist in the country [yuk], who said that the most likely impact of deep-sea disposal of the Brent Spar was 'the death of a number of worms on the sea bed'.

Since I had never spoken to Mr Eggar or, as far as I am aware, to any of his aides, I assume that the 'quote' was taken from one of the newspaper articles. In a couple of steps from my telephone to the House of Commons, my 'millions' had become 'a number'. But there was more. A few weeks later a German friend sent me a cutting from a German newspaper, the *Flensburger Nachrichten*. Here, 'Der Britischen Tiefseeobiologen Tony Rice' says the Spar would cause 'Der Tod von ein paar Würmern auf dem Meeresgrund', that is, the death of *ein paar* (between two and ten according to my friend) of worms. So by now, in only three or four stages my estimate had been cut down by no less than six orders of magnitude – but was still firmly attributed to me!

4 The use and abuse of precedent

1 Shell UK began the removal of the first of its 'small' North Sea structures in this way with the decommissioning of the Leman BK gas compression platform in 1996 (see *Lloyd's List*, 11 July 1998). The Leman platform weighed more than 6,000 tonnes and stood in only 33 metres of water. The topsides, weighing 4,600 tonnes, were removed in October 1996, but technical problems and bad winter weather delayed the rest of the decommissioning process until summer 1997 when the whole of the remaining structure was removed for recycling at the Teesside Environmental Recycling and Reclamation Centre (TERRC). One of the shortlisted options for the disposal of the Brent Spar also involved its recycling at TERRC (see Chapter 8). Subsequently, a further eleven installations have been totally removed, the largest, N NE Frigg FCS, weighing in at almost 13,000 tonnes and standing in 102 metres of water (Rose 1998).

2 In this context, the House of Lords Select Committee on the Decommissioning of Oil and Gas Installations (House of Lords 1996) made the point (4.17) that 'It would be expensive to identify several different sites where the deep sea disposal of oil and gas installations would be permissible, and even more expensive to monitor such sites, which might be scattered over a wide area'. They accordingly recommended 'that if any deep sea disposals are planned in the future the

Government should define the criteria for an ecologically acceptable area in United Kingdom waters within which they will allow the disposal of installations'.

3 'Briefing note on correspondence between Shell UK and UK Government [DTI] regarding the Brent Spar, 1991–5' (Greenpeace, January 1997). From references in the released correspondence, Greenpeace estimated that 'there were at least 122 letters to DTI from Shell in the Shell "Brent Spar Abandonment" file by 19 December 1994. Obviously only a small fraction of this correspondence has been released'.

4 In defence of the DTI, and for that matter of Shell, it is obvious that some anticipation of the outcome would be essential in such a formal and sensitive procedure as that involved in obtaining the necessary UK disposal licence and informing signatories to international agreements. The alternative of leaving all options and possibilities open until the very last moment would lead to chaos. The potential criticism is not that the preferred outcome was decided on early, but simply whether this decision was taken before all the relevant information was available, for example about particular sites, and particularly whether future disposals were already also anticipated without any of the relevant data.

6 The capitulation

1 In view of the hard line taken by the German government it is interesting that the largest intentional disturbance of the deep-sea floor so far undertaken was financed by the German Bundesminister für Forschung und Technologie. The project, called DISCOL (standing for Disturbance and Recolonisation), was organised by one of our colleagues, Professor Hjalmar Thiel, then at the University of Hamburg, and attempted to mimic some of the effects of mining manganese nodules. To do so, in 1989 some 10km^2 of the deep Pacific seafloor about 400 miles southwest of the Galapagos was ploughed up using modified agricultural harrows, intentionally and inevitably causing the deaths of large numbers of seafloor animals, almost certainly many more than would have been killed by a dumped Brent Spar. The disturbed area has been monitored regularly ever since to investigate the rate of recolonisation. One of Greenpeace's experts in Germany, Dr Christian Bussau, based his Ph.D. on the Discol work.

2 Yet another embarassing Shell U-turn made some newspapers in this period. Apparently South African Shell had offered a cash bounty, variously reported as £800 or £1,000, to be paid to any Springbok player for each and every successful tackle on New Zealand's Jonah Lomu in Saturday's rugby World Cup Final. After public outcry the 'offer' was changed to one in which the bounty would be paid into a fund for South Africa's rugby development in black areas for every try scored by the Springboks.

3 The Nisbet and Fowler article prompted indignant protest from Greenpeace – and from many scientists who were horrified at the implied suggestion that the vent sites might be good places to dump

Brent Spars. Unfortunately, the article also threw more confusion than light on the subject because a subsequent letter to *Nature* (Elderfield *et al.* 1995) pointed out significant flaws in the Nisbet and Fowler paper, including the fact that their estimate of the metal output for one of the vents was overcooked by a factor of up to 10,000 times!

7 The aftermath

1 All of these figures are put into a wider context, however, when measured against what could possibly be said to be the biggest media event in recent history, the death of Diana, Princess of Wales. In the month of September 1997, a search for articles in the main quality newspapers and news magazines in the UK regarding Diana discovered a staggering 1,339. This number does not include the copious amounts of column inches and extra 'special edition' memorial issues at the tabloid end of the newspaper market.

8 The way forward

1 Sadly, Professor Henry Charnock F.R.S., who had contributed very significantly to the Group's first report, died before the new study began. Tony Rice decided not to take part in the work of the reconvened Group, but to write his contribution to this book instead.
2 The UK Government issued the relevant licence on 26 August 1998. On 21 August Shell Expro had issued a bulletin giving an update on the progress of the Brent Spar project. It reported that on 12 August the Spar had been moved from Erjford to Vats, a deep-water facility some 60km northeast of Stavanger, where the dismantling operations would be undertaken. It also reported that a new contractor, Heerman Marine, had been appointed to support the dismantling operation and would use a large launch barge instead of the catamaran originally proposed. Heerman would be responsible for the removal of the topsides and the positioning of the ring sections on the seabed at Mekjarvik. The start of this final stage was announced by Shell on 25 November 1998.

9 The Spar legacy

1 The Royal Dutch/Shell Group report even includes 'A personal view', written by John Elkington, Chairman of SustainAbility, the London-based 'values-led' conflict resolution consultancy. John Elkington is well known for his 'triple bottom line' concept of acceptable business activities, that is, taking into more or less equal account the economic, social and environmental aspects and consequences. Since SustainAbility had been appointed by Greenpeace to help develop an overall strategy towards the disposal of offshore structures which was diametrically opposed to that of Shell and the offshore industry in general, a supportive article in Shell's own glossy seems something of a U-turn. But in explaining his contribution, Elkington suggests that the

attitudes of Shell senior management, including Chairman Cor Herkströter and chairman designate Mark Moody-Stuart, indicate a real road-to-Damascus conversion.

2 This point had been made forcibly by the Shepherd Group in only the third paragraph of its first report:

> The environmental impact of the disposal of wastes is a crucial factor which must be taken into account in reaching decisions about disposal options. However, other factors, including social, ethical, aesthetic, legal and economic factors, must be also considered in addition to the scientific evidence'

In its second report, on the final choice after the dialogue process, the Group concluded that the potential environmental impacts of both the original deep-sea disposal option and the final choice would be so small that they could not be used to discriminate between them. Again, the Group suggested that the ultimate decision would have to be taken on other grounds.

10 Epilogue

1 New guidance notes on the BPEO process were issued by the Environment Agency in 1997.

2 In an article in the *Scotsman* on 17 November 1997, reviewing the first of a controversial series of 'anti-green' BBC television programmes, the TV producer Simon Campbell-Jones was quoted as saying

> Friends of the Earth, for example, tend to get their science right but Greenpeace don't really care about the science...the Brent Spar [when Greenpeace was found to have exaggerated the amount of toxic waste the oil platform contained] is just the latest example of them using unsound science.

Not surprisingly, this elicited an irate response from Greenpeace's Peter Melchett (*Scotsman*, 21 November) in which he accepted once again the oil estimate error, but claimed that 'in every other substantial respect, Greenpeace was proved right, while the then government and Shell have been shown to be at best wrong, and at worst, outright disingenuous'.

3 SustainAbility is a London-based 'management consultancy and think tank', founded in 1987 and 'dedicated to promoting the business case for sustainable development'. It aims to help clients, mainly business, to develop solutions to their problems that are socially responsible, environmentally sound and economically viable. Amongst its past contracts are stakeholder dialogues, similar to that proposed by Greenpeace, organised for BP, BP Chemicals, Dow Europe, ICI Polyurethanes, Manweb, Monsanto, Novo Nordisk, Procter and Gamble, and Tioxide.

4 One particularly obvious example was the statement that Dr Martin Angel of the Southampton Oceanography Centre 'had been involved in

work for Shell, preparing the original proposal to dump the Brent Spar'. Not only is this not true, but in the context in which it appears it casts serious doubt on the independence, not only of Martin Angel, but of any other scientist from the same organisation. Despite having this error pointed out to him directly, Chris Rose has left it in.

5 Long before Chris Rose's book appeared, Greenpeace UK had already provided evidence that they had not learned much from the Brent Spar. They are implacably opposed to the recent moves on the part of the offshore oil and gas industry to explore and exploit possible reserves in the deep waters to the northwest of the British Isles. Their basic argument is a very good philosophical one which says that the effect on the earth's climate of using the oil and gas reserves we already know about is likely to be dire. So the last thing we need is to know about additional reserves. But they have hung this excellent argument on the direct risk to the marine environment of the industrial activities, including the threat to 'deep water coral reefs'.

During 1996 a consortium of oil companies, the Atlantic Frontier Environment Forum or AFEN, commissioned an environmental survey of a large area between Shetland and the Faeroes, part of the so-called 'Atlantic Frontier'. A number of organisations participated in this survey using up-to-date acoustic imaging techniques and a variety of sample and data gathering devices. It was a much better designed survey than that conducted at the potential Brent Spar disposal sites, and produced some fascinating results of interest not only for the oil companies but also for deep-sea scientists in general. Amongst other things, and as expected, it revealed the presence of occasional patches of a deep-water coral called *Lophelia pertusa*, which is widespread on the upper continental slope throughout the northwestern European margin. Like warm-water corals, *Lophelia* is a relative of the sea anemones and jellyfishes in which many 'individual' animals live together in a complex interconnected 'colony' and may, on occasion, form massive 'reefs' several kilometres long and tens of metres high. But nothing remotely resembling such reefs has been found in the relevant area and all the experts agree that *Lophelia* is almost certainly restricted here to small isolated clumps. Having been told this emphatically at a public meeting in July 1997, Greenpeace nevertheless issued a press release three days later announcing that it had lodged papers with the High Court in an attempt to block the issue of licences for exploration in the Atlantic Frontier. And a major plank of their case was that 'the Government is failing to protect the cold water coral reef in the Atlantic Frontier region'.

Appendix I International law and agreements

1 International agreements and conventions have to be reflected in national legislation, of course. In the UK the relevant legislation is as follows. The decommissioning of oil platforms and the abandonment of oil and gas fields is governed by the Petroleum Act of 1987. This Act established abandonment standards and approval procedures,

supporting a case-by-case approach to decommissioning. This is the principle that both Shell and the Government of the day made great play of to counter the accusation that the deep-sea disposal of the Brent Spar would act as a precedent for the similar dumping of any number of other structures (see also Chapter 4). On the more general subject of the disposal of wastes at sea, the recommendations of the London Dumping Convention and the 1972 Oslo Convention were initially embodied in the UK's Dumping at Sea Act of 1974 which was subsequently replaced by the Food and Environmental Protection Act of 1985. This Act, known as FEPA, supplemented by the Environmental Protection Act of 1990, is the basic legislation governing the disposal of offshore structures in UK waters. It is admin- istered by the licensing authorities, the Department of Trade and Industry and, for Scottish waters, the Scottish Office Agriculture, Environment and Fisheries Department (SOAEFD) Marine Laboratory in Aberdeen (MLA). In order for them to discharge their licensing duties, both in ensuring compliance with national legislation and with the UK's obligations to the various international agreements and conventions mentioned above, the licensing departments require applicants for licences to comply with the Best Practicable Environmental Option process. Chapter 2 examined the origins and nature of this process, unique to the United Kingdom, and how it worked, or did not work, in the case of the Brent Spar.

Bibliography

ACBE (1997) 'Seventh progress report to and response from the President of the Board of Trade and Secretary of State for the Environment', March, London: Department of the Environment and the Department for Trade and Industry, 49pp.

AURIS (1994) 'Removal and disposal of Brent Spar: a safety and environmental assessment of the options', Aberdeen: AURIS Environmental, 73pp.

Coleman, S. (1997) *Decommissioning Offshore Installations*, London: Financial Times Energy, 214pp.

Corcoran, M. (1995) 'Brent Spar abandonment: a review of the technical case to support deep sea dumping', London: Greenpeace UK.

Det Norske Veritas Industry AS (1995) Brent Spar Disposal Contract C48870/95/DB9. Inventory study (Management summary report, 27pp.; Topside inspection, 39pp.; Tanks and piping, 61pp.).

——(1997) 'Assessment of the proposed options for the disposal of Brent Spar', report no. 970911–0007, 106pp. Prepared for Shell UK Exploration and Production.

ECOS (1996) 'UKCS 17th Offshore Round: environmental screening report, South West Approaches', report 3, carried out by Environmental Services Ltd (ECOS) for the United Kingdom Offshore Operators Association.

Elderfield, H., Schultz, A., James, R., Dickson, P., Mills, R., Cowan, D. and Nesbitt, R. (1995) 'Brent Spar or broken spur?', *Nature*, 376: 208.

Elkington, J. (1997) 'Laying the ghost of the Brent Spar', *Resurgence*, 3, September.

Gage, J. D. and Gordon, J. D. M. (1995) 'Sound bites, science and the Brent Spar: environmental considerations relevant to the deep-sea disposal option', *Marine Pollution Bulletin*, 30, 12, 772–9.

Gao, F. (1997) 'Current issues of international law on offshore abandonment, with special reference to the United Kingdom', *Ocean Development and International Law*, 28, 59–78.

House of Lords Select Committee on Science and Technology (1996) 'Third report: decommissioning of oil and gas installations', 6 March, London: HMSO, 64pp.

Howarth, S. (1997) *A Century in Oil: The Shell Transport and Trading Company 1897–1997*, London: Weidenfeld and Nicholson, 397pp.

Metocean plc (1993) 'Evaluation of environmental aspects of the deep water disposal option', report no. 518, April, prepared for Shell UK.

Natural Environment Research Council (1996) 'Scientific group on decommissioning offshore structures: first report', April, Swindon: NERC, 76pp.

——(1998) 'Scientific group on decommissioning offshore structures: second report', May, Swindon: NERC, 44pp.

Nisbet, E. G. and Fowler, C. M. R. (1995) 'Is metal disposal toxic to deep oceans?', *Nature*, 375: 715 (see also the editorial comment in the same issue).

Rose, C. (1998) *The Turning of the Spar*, London: Greenpeace, 221pp.

Royal Commission on Environmental Pollution (1976) *Fifth Report: Air Pollution Control – an Integrated Approach*, Cmnd 6371, London: HMSO, 130pp. (see also the *Twelfth Report*, 1988).

Royal Dutch/Shell Group (1998) 'Profits and principles: does there have to be a choice?', London: Shell International, 56pp.

Rudall Blanchard Associates Ltd (1994a) 'Brent Spar abandonment BPEO assessment', prepared for Shell UK Exploration and Production Ltd, December, 34pp.

——(1994b) 'Brent Spar abandonment impact hypothesis', prepared for Shell UK Exploration and Production Ltd, December, 40pp.

——(1997) 'Brent Spar supplementary environmental assessment for the deep sea disposal', 26pp.

Shell UK (1987) 'North Sea fields: facts and figures', London: Shell UK Exploration and Production Ltd.

——(1998) 'Report to society', London: Shell UK Ltd, 53pp.

Side, J., Baine, M. and Hayes, K. (1993) 'Current controls for abandonment and disposal of offshore installations at sea', *Marine Policy*, 17 (5) 354–62.

Svitzer Ltd (1996) 'North Atlantic surveys: August 1994–March 1995', report prepared for Shell UK Exploration and Production Ltd.

Taylor, B. G. S. and Turnbull, R. G. H. (1992) 'Development', in William J. Cairns (ed.) *North Sea Oil and the Environment*, London: Elsevier, 697pp.

Index